GALAXIES
AND THE
COSMIC
FRONTIER

GALAXIES AND THE COSMIC FRONTIER

William H. Waller

Paul W. Hodge

HARVARD UNIVERSITY PRESS

CAMBRIDGE, MASSACHUSETTS

LONDON, ENGLAND

2003

Library of Congress Cataloging-in-Publication Data

Waller, William H. (William Howard), 1952–
Galaxies and the cosmic frontier / William H. Waller, Paul W. Hodge.
p. cm.
Includes bibliographical references and index.
ISBN 0-674-01079-5 (alk. paper)
1. Galaxies.   I. Hodge, Paul W.   II. Title.

QB857 .W35 2003
523.1'12—dc21
2002038744

CONTENTS

Contents

PREFACE

The story of our Universe is written in the galaxies. Throughout space and time, galaxies have brought form to the cosmos, light to the void. These myriad constructs of stars, planets, gas, and dust have enchanted and challenged astronomers since 1925, when Edwin Hubble first correctly identified them as autonomous "island universes." Today, the observed structures and motions of galaxies continue to amaze and confound us. And like luminous ships on an ink-black sea, the visible galaxies constitute a mere fraction of what is out there—the ponderous remainder lurking in complete darkness.

In the following chapters, we present the story of galaxies and of their evolution over cosmic time. This astronomical saga is the fifth such accounting published by Harvard University Press, a legacy that harks back to Harlow Shapley's original version of *Galaxies*, first published in 1943. Much has changed in our understanding of galaxies since Shapley's seminal book. Indeed, the pace of discovery has surged forward in astonishing ways since the fourth version was written by Paul Hodge in 1986. Thanks to advanced instruments such as the Hubble Space Telescope and the giant groundbased telescopes atop mountain ridges in Hawaii and Chile, we can now detect galaxies so distant that we are seeing them shortly after their emergence from the din of the Big Bang. Now, more than ever before, we have come to realize that the history of the Universe is embodied in the lives of the galaxies.

To provide both breadth and depth on this enormous topic, we have organized *Galaxies and the Cosmic Frontier* into three parts. In the first, we present a basic overview of galaxies—introducing the elliptical, spiral, and irregular types of galaxies and exploring their corresponding natural histories.

Then we offer some ideas about galaxy formation, along with some delibera-
tions regarding the elusive nature of the dark matter.

In the second part, we present the nearby galaxies up close and personal.
We begin with an insider's tour of our cosmic home, the Milky Way galaxy,
then become acquainted with the numerous dwarfs in our Local Group of
galaxies, probe the exquisite structures and dynamics of the giant spiral and
elliptical galaxies, witness colliding and erupting galaxies, and pay our respects to the most powerful galaxies—the quasars.

In the third and final part, we move beyond our local supercluster of galaxies to contemplate the overall structure, dynamics, and evolution of our
galaxian Universe. As we explore this expanse, we consider the recent discoveries that have revolutionized the study of cosmic evolution. Moving from
maps of vast filaments and voids in the galaxian firmament, to images of
"primeval" galaxies on the cosmic frontier, to hints of a Universe that is expanding ever faster, we end our story of space and time with many loose ends
and exciting possibilities.

Whether starbursts, black holes, hypernovae, dark matter, dark energy, or
cosmic inflation are being discussed, we have focused on the astronomical
observations that inform these astrophysical concepts. We have also emphasized the salient observations that address our cosmic lineage, current state,
and evolving destiny. By hewing closely to the observations while keeping
the big questions in mind, we have endeavored to delineate the history of
our Universe as best we know it today.

We have tried to tell this grand tale in words that can be understood by a
reader with little scientific training. Those who wish to obtain more techni-
cal information are invited to visit our website, <http://cosmos.phy.tufts
.edu/cosmicfrontier>. The website includes tables of physical and astronomical quantities, pertinent links, and updates for further inquiry.

We hope that by writing *Galaxies and the Cosmic Frontier* we have man-
aged to convey some of the spirit and substance of this most excellent
galaxian adventure. We are grateful to the many people who have shared
with us the fruits of their own adventures on the cosmic frontier. In particular, we are indebted to Adam Block, who provided many of the colorful images of galaxy types presented in Plates 1-3; Edwin Krupp, who provided an-
thropological perspectives on the Milky Way; Lori Agan and Leah Staffier,
who crafted many of the figures; our students at Tufts University and the
University of Washington, whose critiques and contributions were of great
help; our colleagues—including David Adler, Chuck Bennett, Tom Dame,
Susan Ernst, Wendy Freedman, Henry Freudenreich, Sol Gittleman, Eva
Grebel, Paul Green, Mike Hudson, Bill Keel, Crystal Martin, Mike Norman,

Ken Olum, Linda Sparke, Ted Stecher, Gregory Taylor, Tony Tyson, and Pat Waller—who provided figures, guidance, and support; and the wonderful editorial team at Harvard University Press, Michael Fisher, Nancy Clemente, and Sara Davis, who coached us through the arduous birthing process. Finally, we salute all private and public benefactors (our readers included) whose interest and support have empowered astronomers worldwide in the pursuit of cosmic discovery.

I

A GALAXY PRIMER

1

GALAXIES
AND THE
UNIVERSE

For as far back as we can tell, our species has sought answers to what lies beyond the horizon. Driven by our inborn curiosity, ambition, and instinct for survival, we have relentlessly ventured forth to explore distant and mysterious lands. Since the last ice age, some 20,000 years ago, our intrepid exploring has led to the expansion of humankind itself. On foot, astride horses, and aboard ships, we have probed most of the secret places on Earth.

More recently, our efforts have opened up vast new realms beyond the confines of our home planet. Beginning with the invention of the telescope in the early 1600s, we have discovered an ever enlarging hierarchy of matter—extending from our Sun-centered system of planets to other stars with planetary systems, to galaxies laden with stars and stellar systems, to groups and clusters of galaxies, to sprawling filaments of galaxies and galaxy clusters at the limits of our visibility. Today, using telescopes, computers, and scribbled notes as our vehicles, we continue to push out to greater and greater distances, searching for the end of space—and the beginning of time.

For most of the last millennium, we equated our Universe with what we today call the Solar System, believing the stars to be merely decorations on a sphere that lay just beyond the planets. By the 1800s, these pinpoints of light were finally recognized as distant suns, so far from us that their radiation takes years to reach the Earth. Space seemed to be sparsely populated by these lonely stars, and scientists argued whether they extended off into the distance forever or whether there was an end to them, beyond which was emptiness. As astronomers explored farther into the depths of space, they did find an end. Our Sun seemed to be one of an immense number of stars that dwelled together in a system they called the Galaxy. Beyond the Galaxy's edge was darkness.

The twentieth century brought about a new discovery: our Galaxy is not the entire Universe. Beyond the most distant stars in the Milky Way are other galaxies like ours, extending out to distances in space that are at the limits of our biggest telescopes. First documented in the 1780s as part of Charles Messier's catalogue of fuzzy objects in the sky, galaxies were finally recognized as "island universes" when their distances were successfully fathomed in the 1920s. Today these magnificent realms constitute one of modern astronomy's most active and exciting topics, and so are accorded the bulk of our attention in this book.

The Scale of Things

Galaxies can be simply described as self-gravitating systems of stars, star clusters, and diffuse clouds of interstellar gas and dust—along with ponderous amounts of dark (nonluminous) matter. All but the last can be directly observed with our groundbased and spaceborne telescopes. Astronomers can now access the full spectrum of radiant energy, from the radio, through the visible, to the highest-energy gamma rays (see Figure 1.1). The dark matter itself can only be inferred from its gravitational effects on the motions of stars and gas within galaxies and on the trajectories of light rays traversing the vast reaches between galaxies (see Chapter 4).

Globular star clusters, such as the Great Cluster in Hercules, qualify as self-gravitating systems of stars (see Figure 1.2). But they are too closely associated with their much larger host galaxies to qualify as galaxies in their own right. We regard globular clusters as dependent members of their host galaxies. On much larger scales, clusters of galaxies are also thought to be self-gravitating systems, but ones in which each cluster member is sufficiently autonomous to be regarded as a separate galaxy. Given these somewhat semantic constraints, galaxies are seen to span sizes of a few hundred light-years to several hundred thousand light-years, a light-year being the distance traveled by light in one year—roughly 6 trillion miles or 10 trillion kilometers. The corresponding contents of these realms amounts to a few million to several trillion Suns, 1 Sun being 300,000 times more massive than the Earth.

The Milky Way, our home Galaxy, is so immense that it takes light over 100,000 years to travel from one side to the other—at the speed of 300,000

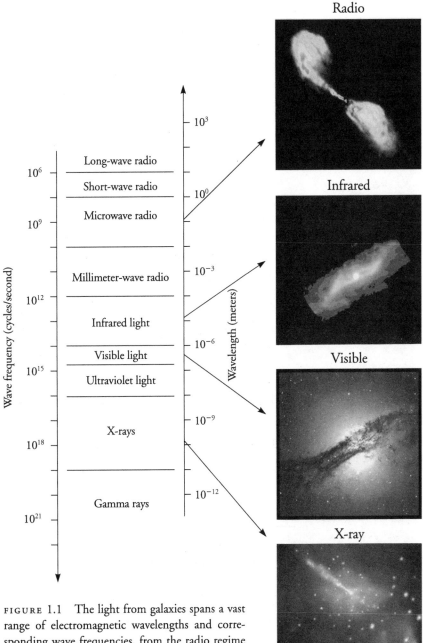

FIGURE 1.1   The light from galaxies spans a vast range of electromagnetic wavelengths and corresponding wave frequencies, from the radio regime (top left) through the visible (middle left) to the highest-energy gamma rays (bottom left). Advances in groundbased and spaceborne telescopes are enabling astronomers to sense the entire spectrum of cosmic radiation with ever increasing acuity. For example, recent multi-wavelength views of the giant elliptical galaxy Centaurus A vividly portray the peculiar nature of this merging system (right). Radio and X-ray images reveal powerful jets emanating from the nucleus. At infrared wavelengths, the inner disk of an embedded spiral galaxy is evident. The visible image highlights the giant elliptical itself, girdled by a band of obscuring dust that was probably once part of the spiral galaxy's outer disk.

100 Light-Years

FIGURE 1.2 Galaxian size scales. From top to bottom: a star cluster in our own Galaxy; a nearby galaxy, seen edge-on with star clusters barely visible in its halo; and a cluster of galaxies.

100,000 Light-Years

1,000,000 Light-Years

kilometers per second, or about 670 million miles per hour. The Earth and Sun lie about 30,000 light-years from the center, or about two-thirds of the way out. If we were to try to signal a hypothetical being who lived near the center of our Galaxy, we would not get a reply until 60,000 years later. To put these distances another way, had we sent the message airmail on the day the Universe began, at the airline speed of about 600 miles per hour, the message would only be about halfway there by now; and the reply would not get back before a total elapsed time of about 70 billion (70×10^9) years.

Some galaxies are much bigger than ours. The largest—such as the giant elliptical galaxy and powerful radio source known as Centaurus A—have total diameters ten times larger than the Milky Way's. Such giants are few and far-flung across the visible Universe. The most numerous galaxies are much smaller than the Milky Way. Dwarf elliptical galaxies, like the tiny one in the constellation Draco, are only about 10,000 (10^4) light-years in diameter. Of course, even these inconspicuous objects are almost unimaginably big. The Draco galaxy might be called a dwarf, but it is more than 100 thousand trillion (10^{17}) miles from one side to the other.

Although there are billions of galaxies, they are not particularly crowded together. The Universe is immense enough to accommodate them comfortably with plenty of space to spare. A typical distance between bright galaxies like the Milky Way is about 5 or 10 million light-years, or roughly 100 times their diameters, with little dwarfs taking up some of the space between.

When their size is taken into consideration, galaxies are much cosier with respect to each other than are the individual stars within them. That is because a star's diameter is infinitesimally small compared to the distance to its nearest neighbor. Our Sun's diameter is about a million miles, whereas the distance to our nearest stellar neighbors (the Alpha Centauri multiple star system) is an incredible 30 million times greater still.

To imagine the spacing between galaxies, think of galaxies scaled down to the size of people. In a typical part of the Universe, the adults (the bright galaxies) would be separated by about 300 feet, with several small children scattered about between them. This part of the Universe would look like a somewhat expanded baseball game, with lots of open space between the players. Only in a few areas, where galaxies congregate in tight clusters, would our scaled-down cosmic scene be as crowded as a city sidewalk, and nowhere would it resemble a cocktail party or a rush-hour subway car. By comparison, if the stars in a typical galaxy were scaled down to the size of people, the countryside would indeed be a lonely place. Typically, your nearest neighbor would live about 60,000 miles away, or about one-fourth the distance to the Moon.

These examples should help to illustrate that galaxies are sparsely strewn about the Universe, and that they contain mostly empty space. Even when we take into account the clouds of gas and dust that lie between the stars within a galaxy, the densities of matter are still extremely small. Our Universe of galaxies is huge, and it is very nearly empty.

Kinds of Galaxies

Like trees, galaxies come in an amazing variety of sizes, shapes, hues, and constituent structures (see Plates 1–3). And like botanists, astronomers have endeavored to make some sense out of this diversity by devising classifications according to common themes. In so doing, astronomers have developed alphanumeric codes that describe galaxies with a minimum of characters. These classifications have also led to critical insights regarding the actual nature of the galaxies themselves.

Edwin Hubble was the first to classify the galaxies—shortly after determining a distance to the great spiral nebula in Andromeda (M31, number 31 in Messier's seminal catalogue) that was 50 times farther than anything else in our Milky Way galaxy. Beginning in the 1920s, Hubble divided the galaxies into three main categories according to their shapes: elliptical galaxies, spiral galaxies, and irregular galaxies, abbreviated E, S, and Irr. He then proceeded to establish a sequence incorporating and relating these categories (see Figure 1.3). Elaborations of Hubble's original sequence continue to aid astronomers who are researching the origins and lives of galaxies.

Elliptical galaxies are characterized by their overall elliptical shape and show no further structure other than a general decrease in brightness outward from a central point (see Plate 1). The precipitous rate at which the brightness decreases outward follows a simple mathematical law, also discovered by Hubble, which goes roughly as the inverse square of the distance, such that a doubling of the radial distance would diminish the brightness by a factor of four.

To put this inverse-square principle into more graphic form, you can draw a line connecting all the points where the brightness of a galaxy appears to be the same, and then other lines for other levels of brightness (analogous to the lines of constant elevation on a topographic map). In an elliptical galaxy, the resulting lines of constant brightness—known as isophotes—would form a series of ellipses, all with roughly the same center and roughly the same shape (see Figure 1.4). If the isophotes are set at equal intervals of brightness, then the inverse-square decline in brightness would be manifested in ever increasing spaces between isophotes. The observed falloff in brightness is so ex-

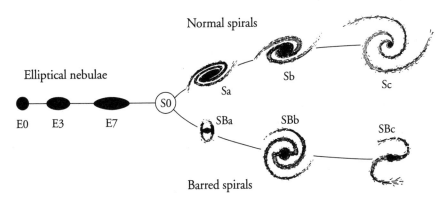

Normal spirals

Elliptical nebulae

E0   E3   E7

S0

Sa

Sb

Sc

SBa

SBb

SBc

Barred spirals

FIGURE 1.3 Edwin Hubble's original "tuning fork" diagram of galaxy types. Hubble sequenced the galaxies from the "early-type" ellipticals (E0 to E7) to the "later-type" normal (Sa to Sc) and barred (SBa to SBc) spirals, thinking that galaxies slowly evolved over time from ellipticals to spirals. We now think that galaxy type is for the most part determined from initial conditions, with any subsequent evolution resulting (mostly) from interactions between galaxies and tending in the opposite direction, from spiral to elliptical. The forking accounts for unbarred (S) and barred (SB) spiral galaxies.

treme, however, that astronomers usually choose to logarithmically compress the wide range of brightness into intervals of equal magnitudes; an interval of one magnitude corresponds to a factor of 2.5 in brightness, and so a five-magnitude interval would amount to a brightness factor of $(2.5)^5$, or 100. The inverse-square decline is then visible as a nesting of more regularly spaced contours with some bunching toward the center (see Figure 1.4).

Subclasses of elliptical galaxies are designated by a number following the letter E that indicates the degree of ellipticity. The number, n, is defined as n = 10 (1 − b/a), where b/a is the ratio between the short and long axes of any one of the galaxy's elliptical isophotes. Thus an elliptical galaxy that has a circular outline will be classified as an E0, and a highly eccentric elliptical galaxy with a ratio between axes of 1/2 will be classified as E5 or so (see Plate 1).

Spiral galaxies feature flat disks containing spiral arrangements of enhanced starlight. At the center of most spirals, a nuclear bulge can be seen protruding above and below the disk. Inside the bulge, a distinct bright nucleus is often evident.

Ideal spirals have two arms emanating either from the nucleus itself or from the ends of a bar centered on the nucleus. These shapes have led to two principal subcategories, normal spirals (S) and barred spirals (SB). Normal spirals outnumber barred spirals by a factor of three or so. Further subdivisions are determined by three criteria: (1) the relative prominence of the gal-

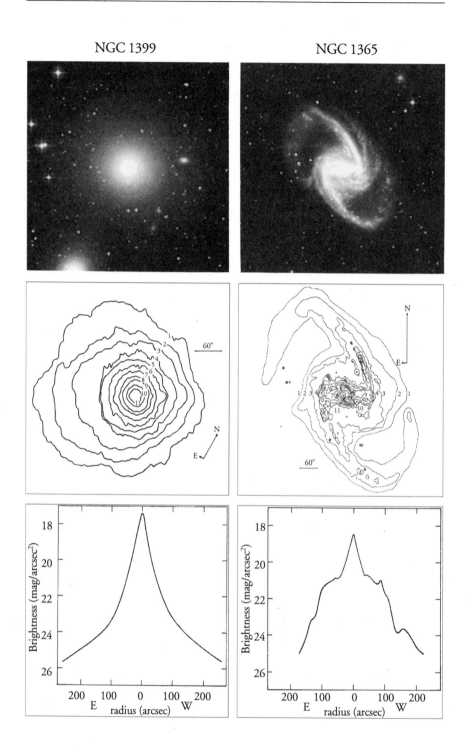

FIGURE 1.4 (facing page) Comparison of an elliptical galaxy and a barred-spiral galaxy. NGC 1399 is a giant E0 galaxy in the constellation of Fornax. Its contour diagram of surface brightness shows regularly spaced elliptical isophotes. Its steeply falling profile of surface brightness (in logarithmically compressed units of magnitudes per square arcsecond of angular area on the sky) is typical for an elliptical galaxy. The center itself is extraordinarily luminous, a condition that suggests an extreme concentration of stars and perhaps a super-massive black hole within. NGC 1365, also in Fornax, shows a strong central bar whose extremities connect with two symmetric outer spiral arms. The east-west profile of surface brightness reveals a centrally peaked bulge and an exponential disk component, whose light declines less rapidly with distance from the center.

axy's nuclear bulge compared to its disk component; (2) the tightness with which the arms are wound; and (3) the resolution of the arms into patches of luminous stars.

The Sa (and SBa) galaxies have very prominent central bulges and tightly wound (nearly circular) spiral arms that are continuous and appear smooth rather than patchy (see Plate 1). The Sb (and SBb) galaxies have somewhat weaker bulge components and more loosely wound spiral arms that are re-solved to some extent into patches of starlight (see Plate 2). Sc galaxies (and their barred counterparts) have inconspicuous bulges with open and frag-mented spiral arms that are highly patchy. In SBc galaxies, even the bar is re-solved into patchiness. In Sd galaxies (an extension of Hubble's original se-quence), the bulges may be entirely absent, with the spiral structure barely traceable.

For most spiral galaxies, the light in the disk decreases with the radius in an exponential manner, whereby the brightness dims by the same fraction over equal intervals of radius. A pure exponential disk would yield a contour diagram of equally spaced magnitude intervals. By contrast, the nuclear bulge component has many of the characteristics of an elliptical galaxy, in-cluding a dense bunching of the central contours. In fact, the centrally peaked brightness law that Hubble found to apply to elliptical galaxies also applies to the central regions of spiral galaxies, and so the central bulge is sometimes called the "elliptical component" (see Figure 1.4).

Irregular galaxies make up the category into which Hubble placed all those that did not fit into the elliptical or spiral classes (see Plate 3). Most ir-regular galaxies are rather similar to each other in appearance, being ex-tremely patchy so that individual bright stars and regions of hot, luminous gas can be discerned. Some irregular galaxies have conspicuous bars, and many of them have what appear to be patches of structure that suggest fragments of spiral arms. Few of the irregulars have central bulges, and almost none of them have distinct nuclei.

Hubble realized that irregular galaxies of this variegated type (which he called Irr I) seem to be extreme extensions of the spiral galaxy classification, where the arms are so patchy and disjointed that they cannot be justifiably described as spirals (see Plate 3). With improved distance determinations in hand, we now know that these galaxies are invariably much smaller and less massive than their spiral cousins.

Other peculiar galaxies, lumped among Hubble's original class of irregulars, do not seem to be related to the more common irregular objects, usually because of the presence of dust, distorted shapes, or other anomalous features. These were designated Irr II, but in subsequent revisions of the Hubble classifications many of these strange-looking irregular galaxies were reclassified into "S0," "Amorphous," and other subclasses. For example, galaxies with flat disks, like the spirals, but no spiral arms are termed S0 galaxies. These puzzling lenslike (or "lenticular") systems are now regarded as "early-type" galaxies with traits shared by both the ellipticals and the Sa-type spirals (see Plate 1). A few galaxies still remain outside the classification, and many of them have since been discovered to be either interacting pairs or subject to some other violent event of internal origin (see Plate 3).

Last, but certainly not least in number, there are the dwarf galaxies. Compared to the giant spirals and ellipticals, these galactic lightweights occupy a distinct second tier of galaxy types. To date, astronomers have found dwarf ellipticals, dwarf irregulars, and dwarf spheroidals—the last often appearing as little more than a haze of dim stars amidst the brighter foreground stars of our Milky Way galaxy. Dwarf spirals have yet to be found, which suggests that the generation and sustenance of spiral structure require more massive host galaxies (see Chapters 2 and 9).

Astronomers have found a bounty of dwarf galaxies around the Milky Way and Andromeda galaxies—the two giant systems that dominate our Local Group of galaxies. No doubt, other groups of galaxies are similarly replete with dim dwarfs. Despite their great numbers, most dwarf galaxies outside our Local Group are too faint to be studied in any detail. Consequently, great pains have been taken to understand those nearby dwarfs that we can observe in depth (see Chapter 8).

Astronomers are especially interested in the dwarfs, because they provide important clues to the nature of the galaxian Universe shortly after its emergence from the Big Bang—the hypothetical moment of cosmic creation some 15 billion years ago. Dwarf-size protogalaxies may have been the first self-gravitating large objects to form out of the cooling plasma of the prime-

Chapter 15). Also, dwarf galaxies are relatively unevolved in
ical makeup of the stars that have formed in them. The
of special interest, because most of their gaseous stores
rted into stars—yet another indication of nearly primor-

What Are We Missing?

e attracted to what we can see. This selection effect
ystem that is strongly biased toward the luminous
laxies. The many dim dwarfs in our Local Group
here being a lot of galaxian matter that is proba-

bact that they are indistinguishable from the
alaxy. Fortunately, their spectra often reveal
centrally concentrated galaxies at great dis-
shifted spectral features, produced by light
greater (redder) wavelengths, provide reli-
ources are participating in the overall ex-
ce are located well beyond the Milky Way.
Man   ait pointlike sources with highly redshifted spectra, were
discovered in this manner (see Chapter 11).

Other galaxies are so diffuse that they barely show above the background
level of the night sky. These so-called low surface–brightness (LSB) galaxies
include both dwarfs and unusually dim giants. Enough LSB galaxies have
been discovered to suggest that they represent an important galaxian population whose origin and evolution need explaining.

Classification Schemes

Since Hubble's pioneering work in the 1920s, there have been several inno-
vative attempts to codify the remarkable variations among galaxies. Each of
these classification schemes incorporates Hubble's original sequence of galaxy
shapes while highlighting other salient features.

In the 1950s, Gerard de Vaucouleurs developed a three-dimensional classification system (see Figure 1.5). Along the "major axis" of this system, a
variant of Hubble's basic sequence runs from E0 ellipticals through the "lenticular" (S0) galaxies and Sa-Sc spirals, with an additional Sd class, and a
Magellanic irregular (Sm) class (named after the Milky Way's most prominent companion galaxies, the Large and Small Magellanic Clouds) that tran-

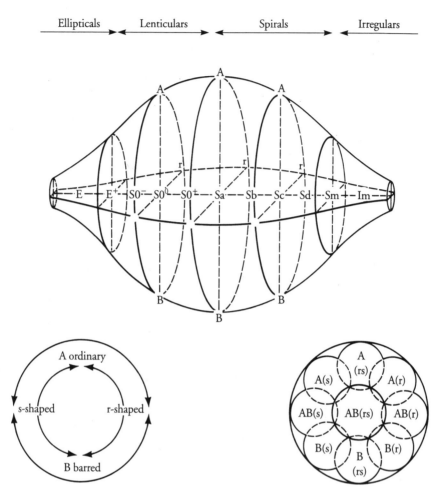

FIGURE 1.5 The three-dimensional classification system of Gerard de Vaucouleurs organizes galaxies according to their Hubble type, bar strength, and inner structure. The primary classification parameter is the Hubble type (E-Im), which follows the degree of central concentration and the tightness of the spiral winding. The bar strength index ranges from negligible (A), through intermediate (AB), to prominent (B). The inner structure parameter ranges from ringlike (r), through intermediate (rs), to spiral-like (s). The variations in bar strength and inner structure are most pronounced in galaxies with Hubble types near Sa.

sitions into the irregular category (Im). Instead of making a separate branch for the barred spirals, as in Hubble's "tuning fork" diagram, de Vaucouleurs created a perpendicular axis that indicates the bar strength—from A (unbarred), through AB (ovoid), to B (barred). Perpendicular to these two axes, he created yet another axis that codifies the structure of the inner disk—from

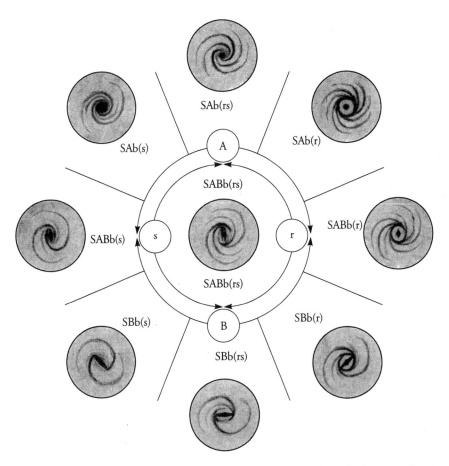

FIGURE 1.6 This galaxy mandala represents a cross-section through de Vaucouleurs' three-dimensional system of galaxy classification. It is centered on the intermediate Hubble type Sb, with bar strength increasing toward the bottom and ringlike structure increasing toward the right. The central SABb(rs) class comes close to what our Milky Way galaxy is thought to resemble.

spiral-like (s), through spiral and ringlike (rs), to ringlike (r). This third dimension highlights important morphological clues to some of the most fascinating dynamics taking place in spiral galaxies (see Chapters 2 and 9).

The "galaxy mandala" shown in Figure 1.6 is actually a cross-section through the de Vaucouleurs classification system, centered on the Sb type. The central galaxy in the mandala has the intermediate classification of SABb(rs), and is thought to be a close approximation to what our Milky Way galaxy looks like (see Chapter 6). More recent revisions of the de Vaucouleurs system recognize the presence of outer rings or pseudo-rings that have been revealed in long-exposure images, by respectively adding an R or R′ prefix to the other characters.

Although such detail can be considered overkill when astronomers are trying to describe poorly resolved galaxies at great distances, the de Vaucouleurs system economically encodes the greatest amount of information for galaxies in the local Universe (within a few hundred million light-years or so). Indeed, the *Third Reference Catalogue of Bright Galaxies (RC3)*—one of the greatest compilations of galaxies to date—incorporates an alphanumeric version of the de Vaucouleurs classification system. Consequently, astronomers often refer to the de Vaucouleurs classifications when researching and writing about galaxies.

In 1960 the Canadian astronomer Sidney van den Bergh found a series of morphological clues in the photographs of spiral galaxies that could be used to estimate the absolute luminosities of those galaxies, the luminosity being their intrinsic radiating power. Like the de Vaucouleurs system, van den Bergh's luminosity classification complements Hubble's Sa—Sc sequence of decreasing bulge prominence, increasing arm openness, and patchier arm texture. A spiral galaxy of a given Hubble type, such as an Sc galaxy, which has a small bulge and relatively open spiral arms, could be of any van den Bergh class from I to V. The smaller number would indicate longer spiral arms of higher "quality" (contrast and form)—and would also predict a larger and intrinsically more luminous galaxy.

Calibrating his luminosity classes by examining galaxies of known luminosity, van den Bergh showed that the full range of luminosity classes represents a factor of about 400 in actual luminosity, or a difference of 6.5 magnitudes in the logarithmic system of absolute magnitudes, where each increment of one magnitude corresponds to a factor of 2.5 in luminosity. Further tests of van den Bergh's admittedly qualitative classification scheme indicate that it can be used as a fairly reliable measure of a spiral galaxy's intrinsic luminosity (see Figure 1.7).

Closely following in the footsteps of Edwin Hubble, Allan Sandage published in 1961 the renowned *Hubble Atlas of Galaxies*. Therein, he described the Hubble sequence in terms of archetype galaxies, using stunning pictures obtained with the 100-inch (2.5 meter) telescope on Mount Wilson and the 200-inch (5-meter) telescope atop Mount Palomar in California. Later, he set about extending Hubble's classification system and applying it to a greater number of nearby galaxies. Together with Gustav Tammann, he published *A Revised Shapley-Ames Catalog of Bright Galaxies* in 1981, wherein 1,246 galaxies are classified.

In 1994, Sandage and John Bedke published *The Carnegie Atlas of Gal-*

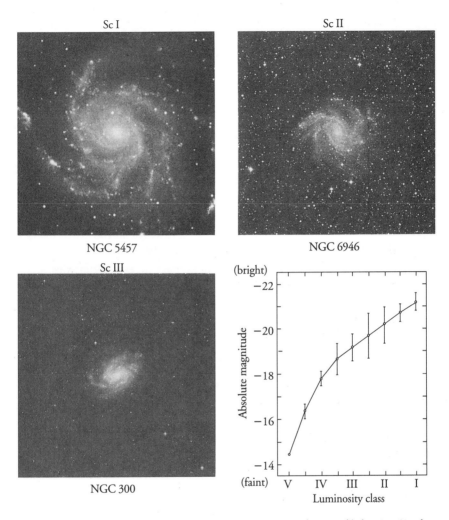

FIGURE 1.7 The clearly delineated relation between van den Bergh's luminosity class (I-V) and a galaxy's intrinsic luminosity, expressed in absolute magnitudes; the more luminous galaxies have the lower (more negative) absolute magnitudes. This relation indicates that the "quality" and length of a galaxy's spiral arms can be used to quantitatively diagnose that galaxy's overall luminosity. The three galaxies of luminosity class I, II, and III shown here have been scaled according to their actual sizes, to demonstrate the differences in size (and total luminosity) between the luminosity classes.

axies—a two-volume tome that sets the standard for galaxy atlases. In it, Hubble's original classification system is extended to include the Sd, Sm, Im, and Amorphous types, the ring (r) and spiral (s) descriptors of de Vaucouleurs, and the luminosity classes of van den Bergh. For example, the Great Nebula in Andromeda (M31) is classified as an SbI-II galaxy, the smaller and fainter Pinwheel galaxy (M33) is classified Sc(s)II-III, and the Large and Small Magellanic Clouds are respectively given SBmIII and ImIV-V classifications. The so-called Hubble-Sandage system incorporates many of the features of the other systems, while not becoming overly detailed. For these reasons, we will make use of the Hubble-Sandage system in subsequent descriptions of nearby galaxies.

Typecasting Galaxies on the Cosmic Frontier

All of the classification schemes described above fail when one tries to describe extremely distant galaxies. At distances exceeding 5 billion light-years, even the most colossal galaxies appear as blurry smudges. Moreover, the light from these remote sources is subject to severe redshifting—as the electromagnetic waves stretch along with the expanding Universe. What was originally ultraviolet emission is redshifted into the visible spectrum, while the visible emission gets redshifted into the infrared. Consequently, we tend to see only the ultraviolet light from distant galaxies—and, as we will find in later chapters, the shapes of galaxies at ultraviolet wavelengths can differ radically from their visible forms.

To overcome this severe wavelength dependence, researchers of distant galaxies are exploring new ways of classifying what they see. Some astronomers have been "benchmarking" the ultraviolet-emitting structures of nearby galaxies in an effort to interpret deep images of highly redshifted galaxies. Meanwhile, theorists are "training" computers to make the interpretive transformations from what is observed to what would be observed in the absence of redshift.

Others are investigating the degree of central concentration in galaxies, an idea first suggested by William Morgan in the 1950s, when he noted that the elliptical-spiral-irregular Hubble sequence follows a sequence of ever decreasing central concentration. Further study of Morgan's system has shown that his sequence of decreasing central concentration follows a parallel sequence of spectral features similar to those respectively emitted by low-, medium-, and high-mass stars. Therefore, it may be possible to classify distant galaxies on the basis of their spectral features and constituent stars rather than on the basis of their poorly resolved and wavelength-dependent structures.

With the next generation of space telescopes, astronomers will be able to obtain deep high-resolution images of distant galaxies at far-infrared wavelengths. By keeping up with the high redshifts associated with these remote realms, we should then know what kinds of galaxies prevailed at distances of

5–15 billion light-years. And because the light from these galaxies has taken 5–15 billion years to reach us, we will also learn what galaxies were like 5–15 billion years ago, when the Universe was much younger. No doubt, our explorations on the "cosmic frontier" will yield new ways of categorizing galaxies—and new insights into the lives of galaxies over cosmic time.

Explaining Galaxy Types

Ever since Hubble, astronomers have tried to uncover the processes that give galaxies their shape. Some of the early theories attributed galactic shapes to an evolutionary progression. Galaxies were thought to start out as one type that then gradually evolved into the other types. One idea was that they began as elliptical galaxies, developed spiral structure, and then disintegrated into irregular chaos. Alternatively, the process might have worked the opposite way: galaxies began in disorder, gradually wound themselves up as spirals, and ended their evolution with the pure symmetry and simplicity of ellipticals.

Both theories were based on the premise that a galaxy's type reflects its age. Neither notion had any detailed physics to back it up, and after many years of research, both turned out to be mostly wrong. Once astronomers had learned enough about the evolution of stars to be able to measure stellar ages (in the 1950s), they found that all types of galaxies are about the same age. Virtually all galaxies have at least some stars that are several billion years old, an indication that neither elliptical nor irregular galaxies are younger than spirals.

It is true, though, that elliptical galaxies are made up almost exclusively of old stars, while the other Hubble types contain relatively more young stars. Thus the Hubble sequence of shapes does have something to do with the ages. Apparently, the rate of new star formation that has gone on since a galaxy's birth, and hence the distribution of stellar ages within that galaxy, is related to its shape. Elliptical galaxies have had little star-forming activity since their birth and so have few young stars; Sa galaxies are still forming stars, but slowly; Sb's have more star formation going on; Sc's are rife with activity; and irregular galaxies are the most active of all.

These facts have led to the idea that the Hubble sequence is a conservation sequence. The irregulars have conserved more of their gas and dust for star-forming, while the ellipticals used practically all of it up in one initial burst. But how did this difference lead to such different forms? This point will be taken up in Chapter 5, which summarizes our understanding of galactic evolution. Current ideas, now fully backed up with many kinds of evidence, in-dicate that the two most important things responsible for a galaxy's form are its initial conditions (mass and angular momentum) and its environment (whether it has close companions or is a member of a cluster). In this sense, galaxies are a bit like people: their personalities are shaped both by their he-redity and by the company they have kept.

2

FORM AND
FUNCTION

Galaxies are among nature's most beautifully formed objects. Deep photographs of galaxies, taken with the most advanced telescopes, reveal enthralling structures on almost every scale. Even when glimpsed through the eyepiece of a small telescope, these ghostly realms can mystify and inspire.

Why do immense spiral arms wind around the centers of some galaxies? Why do bright bars of starlight span the faces of other galaxies? How can we explain the smooth and symmetric purity of yet other galaxies?

The best approach to answering these questions has involved measuring the luminous structures of galaxies, gauging their internal motions, and then comparing the observations with physical models. In this way, astronomers have been able to test different assumptions about how galaxies formed and how they currently function. This chapter follows that plan; first we compare the "photometric" and "kinematic" properties of galaxies of various types, and then we examine theoretical models to see how well they fit the observations.

How to Build an Elliptical Galaxy

The simplest-looking galaxies are the ellipticals: smooth, uniform in color, and symmetric. The near perfection of their structure hints at their essential simplicity, and it is in some ways true that ellipticals have been easier to measure and fit into theoretical models than have their more complex cousins.

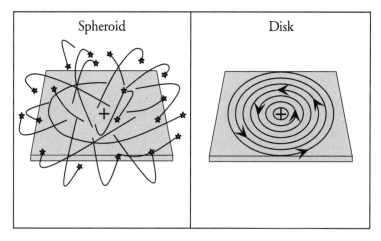

FIGURE 2.1 Comparison of orbital motions in elliptical and spiral galaxies. Ellipticals and the bulges of spiral galaxies (left) contain stars in plunging elliptical orbits that collectively resemble a haphazard swarm. The disks of spiral galaxies (right) contain stars and gas in nearly circular co-planar orbits. The innermost stars and gas, having less distance to travel around the disk, continuously overtake their outer counterparts. This differential rotation, or "shearing," produces the winding dilemma visualized in Figure 2.2.

Let's look at the actual structure of a typical elliptical galaxy. The *New General Catalogue of Nebulae and Clusters of Stars*, first published in 1888, contains thousands of objects that we today recognize as galaxies. Among them, NGC 1399 is an excellent example of the elliptical genre (see Figure 1.4). This E0-type galaxy has a bright nucleus at its center, which is surrounded by diffuse light that decreases steadily outward. As with all ellipticals, the rate at which the brightness goes down can be quantified via simple mathematical expressions such as the inverse-square falloff with radius discussed in Chapter 1. Also the outline of the galaxy—its shape—remains nearly the same at all light levels. Its isophotes are all, as nearly as we can tell, perfect ellipses, centered precisely on the galaxy's nucleus. Their major axes line up in nearly the same direction, and the ratios of major to minor axes are all almost identical.

This basic simplicity is consistent with the idea that elliptical galaxies are governed by very few forces. The stellar motions appear to be well mixed and smooth—consisting of myriad plunging elliptical orbits with no common direction and hence no overall rotation (see Figure 2.1). Nothing but gravity influences the stellar arrangement, and little in the way of ongoing star formation mars the smooth appearance of these galaxies.

When Hubble first pointed out most of these facts, he showed that the structure of an elliptical was akin to that of a simple sphere of gas, arranged by gravity alone and full of identical particles of similar speed. Such an ensemble of stars would be regarded as having a "temperature" corresponding to the degree of motion that the stars share. Models of elliptical galaxies that incorporate this analogy with a constant-temperature gas are collectively known as isothermal models.

To build an isothermal elliptical galaxy out of stars, place the stars near one another in space and let their motions all become similar. Do not give them any global systematic motions, such as rotation, and make sure that you have chosen quiet, well-mannered stars that will not erupt, eject matter, or otherwise disturb the stellar mélange.

It is not necessary to arrange them in a perfect sphere at first. You can let them out of a rectangular box, for example, and just wait awhile; they will eventually arrange themselves into a spheroidal swarm. Gravity acts with spherical symmetry, so if your box of stars has only gravity governing it, it will smooth out, lose its rough edges, and become a nice elliptical galaxy.

Real elliptical galaxies are not perfect spheres, of course. The isophotes of NGC 1399, for example, are ellipses rather than circles, and their axial ratios vary slightly at different distances from the center, making the isophotes less circular in the outer parts. Their alignment also twists a little. All these imperfections tell us that the simple model of elliptical galaxies is not quite right. Some history and special circumstances must have influenced the stellar orbits in a perceptible way. Perhaps some residual rotation is involved; perhaps tidal action from neighboring galaxies has been a factor; or maybe we are seeing the effects of special initial conditions so strong that gravity has not had time to wipe them out completely.

An important clue to the oddness of ellipticals comes from recent observations that show the ellipticity to be essentially unrelated to the degree of rotation. Something else must be responsible for flattening out the distribution of stars in E5-E7–type galaxies. What that is might be related to preferred "families" of orbits that get established in these deceivingly simple systems. Some of these orbital families look like boxes full of plunging elliptical orbits rather than ellipsoidal swarms. Others look like doughnuts, where the stars perform loop-the-loops as they orbit around the galaxy's center.

Another clue to the form and function of ellipticals is the observed steep decrease in brightness with increasing distance from the center. As noted by Hubble, the brightness declines roughly as the "inverse square" of the projected radius—($I(R) \propto 1/R^2$). A deprojection of the observed intensities on the sky into the more physical units of luminosity per unit of volume yields an even steeper decrease with radius. Instead of an inverse-square relation, one obtains an inverse-cube dependence, such that a doubling of the radius yields an eightfold decrease in density.

Such a steep falloff in the stellar density is not predicted by the simple model of self-gravitating "isothermal" stars. Indeed, one is lucky to manage an inverse-square decline. Two processes have been proposed to explain the steeper dependence that is observed: evaporation of the outer stars and concurrent condensation of the innermost stars—the combined processes being known as dynamical relaxation.

The evaporative process can be catalyzed by close tidal interactions between an elliptical and some other galaxy. Some of these interactions may lead to complete mergers of the two systems. Alternatively, encounters between stars and star clusters within the elliptical galaxy itself can redistribute the energies, so that the most energetic stars evaporate. In both of these scenarios, the central condensation ensures that the overall energy of the system remains conserved.

The collapsed core also provides favorable conditions for spawning a super-massive black hole and its subsequent pyrotechnics. Black holes of the galactic variety are like their stellar counterparts, but are much bigger. Both involve concentrations of mass so intense that not even light can escape from their gravitational clutches. Unlike stellar black holes, the black holes that are thought to dwell in the cores of giant elliptical galaxies have masses that range from millions to billions of Suns. Anything that falls into these dark chasms of distorted space and time will release its gravitational energy in wild ways. Indeed, recent observations of giant ellipticals have revealed tremendous outpourings of matter and energy from their dense cores, along with rapid rotation of their innermost gas—all of which implicates super-massive black holes at work in these seemingly benign systems (see Chapter 11).

Disks and Bulges

Unlike ellipticals, spiral galaxies are characterized by both bulge and disk components. The bulge component is like the ellipticals in form and color. It is thought to be populated mostly by old stars in haphazard orbits. By contrast, the disk component contains both young and old stars along with clouds of gas and dust—all in co-planar circular orbits (see Figure 2.1). The spiral arms that highlight the disk component are minor in terms of the number of stars they contain, but they are important because of the exquisite dynamics that they trace in the disk. In a similar sense, the eyes of the human face are a small part of the body, yet they command our attention and reveal a great deal about what is going on inside.

The disk of a typical spiral galaxy is very flat and thin. The spiral galaxies that we happen to see edge-on tell us that the total thickness of a typical disk is about 1/10 to 1/50 its diameter. For our own Galaxy, where we can count stars in the disk and thereby measure its thickness, we find that the stars thin out rapidly—becoming quite sparse at about 1,500 light-years above and below the plane. Were our Milky Way galaxy the same size as a compact disk (CD), its stellar disk would be the equivalent of three CDs stacked together.

The thinness of the disk is especially pronounced for the youngest stars and for the raw materials (gas and dust) that lie waiting to form future stars. This "active" component—in a galaxy the diameter of a CD—would be no thicker than a single CD. As seen in some edge-on spirals, the pencil-thin lanes of gas and obscuring dust delineate the very middle of the disk, whereas the oldest stars in a disk are arrayed in a much thicker band. From this age-sensitive variation in thickness, we surmise that stars originally form in a thin disk and are then jostled by their neighbors into greater excursions above and below the midplane.

Measures of the light from the disks of spiral galaxies show an important commonality that is well documented but not yet satisfactorily explained. The brightness decreases outward in a fairly regular and universal mathematical pattern, but one that tends to be less drastic than that of elliptical galaxies (see Figure 1.4). All disks follow this "exponential" falloff in brightness, from those of tiny dwarf galaxies like GR8 to those of supergiant spirals like M101. As Sidney van den Bergh noted in 1998, "it is not yet clear why Nature appears to be so inordinately fond of exponential disks!"

Computer-generated models of rapidly rotating stellar systems make what we see of galaxian disks seem quite natural. Consider the elliptical galaxy that was described above. If its protogalactic cloud of gas could be set to spinning rapidly before most of the stars were formed, it would resemble the disk of a spiral galaxy. First, the rotation would induce an overall flattening of the collapsing cloud. Second, dissipative collisions between gas particles as they traverse the midplane would nullify all motions except the shared rotation. What remains would be a thin disk of material in co-planar orbits. Thus it appears that the essential structural difference between elliptical galaxies and spirals is the amount of initial rotation.

Then where does the bulge come from? If a rapidly rotating protogalactic cloud produces a disk, while a slowly rotating or nonrotating one produces an elliptical galaxy, what are those fat, ellipsoidal bulges doing at the centers of spirals? They have most of the structural properties of elliptical galaxies: regular isophotes, old stars, considerable thickness, and rapidly decreasing radial light profiles. The answer seems to lie in the important fact that gas behaves very differently from stars. A cloud of gas can get rid of energy fairly easily by virtue of its particles colliding with each other, heating up, and radiating the energy away. As it does so, a rotating gas cloud will rapidly flatten to a disk. But if at any time the gas starts to condense into stars, the situation will change. Stars do not collide as atoms of gas do. They are much too small relative to the distances between them to hit each other. Because they do not heat up as a result of collisions, stars do not dissipate energy efficiently, and hence they do not collapse to a plane. So if stars start to form early—and they will do so first in the central regions where the densities are greatest—then they will maintain their plunging orbits and so form a big thick central bulge.

In the Milky Way, for instance, some of the oldest stars are found in the central bulge. These stars represent a primeval epoch when much of the Galaxy was in free-fall, building its stellar bulge out of the infalling gas. Whatever gas remained was subject to the combined effects of rotation and dissipative collapse, the result being a thin disk, where the gas and consequent stars all rotate together.

This thin, flat disk has become the seat of most subsequent activity in our Galaxy. Virtually all of the stars that we see in the sky, the giant molecular clouds that are incubating new stars, the irradiated and windblown nebulae resulting from the most powerful stars, and the large-scale spiral patterns that orchestrate new starbirth activity have all evolved here, in an intricate interplay that currently challenges the best of astronomers' theoretical efforts.

The Riddle of Spiral Structure

Spiral galaxies would not look very interesting without their spiral structure; certainly they would not be spiral galaxies, but the point is a little more subtle than that. If a galaxian disk forms because rotation forces the protogalactic gas to collapse to a plane, then the spiral structure in the disk is seemingly a natural result, much like the pattern of cream stirred into a cup of coffee, or storms over the Caribbean, or water going down the drain. These more familiar patterns are not strictly analogous to the case of a galaxy, but the connections are illustrative: where there is rotation there is likely to be spiral structure. For many years, therefore, astronomers were not particularly concerned about the fact that many galaxies had a spiral shape—it seemed quite natural.

The first serious difficulty arose when someone thought to ask: How long can a spiral arm last in a galaxy? We know the rotation periods of galaxies to be typically a few hundreds of millions of years for stars that lie at about the equivalent of the Sun's distance from the nucleus. We also know that galaxies undergo differential rotation, such that the inner parts of a galaxy rotate

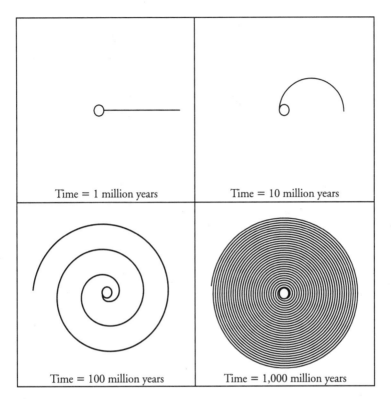

Time = 1 million years	Time = 10 million years
Time = 100 million years	Time = 1,000 million years

FIGURE 2.2 The winding dilemma in disk galaxies. The differential rotation in galaxian disks will shear an initially radial structure into a spiral arm that winds up like a watch spring over time. Because the *observed* number of spiral windings in galaxies is never more than two, there must be some mechanism that prevents the expected large number of windings.

faster than the outer parts. Finally, we know the ages of nearby galaxies to be about 10 or so billion years.

If the spiral structure results from the differential rotation, then the arms should gradually wind up into a spiral pattern (see Figure 2.2). But the number of windings should be very large for a galaxy as old as those around us, about as many windings as the age (10^{10} years) divided by the mean period of rotation (10^8 years), which would be about a 100 or so. Actual spiral galaxies, at least those that have clear, continuous spiral arms, show only one or two windings. And therein lies the puzzle. Do spiral arms "freeze" somehow, so that they can persist? Or do they keep winding up to oblivion, to be succeeded by new ones? Or is there some way in which they do not partake in the general rotation of the stars and gas, so that they rotate more slowly? This puzzle is well known to astronomers as the winding dilemma.

The problem is not that we cannot think of a way to make a spiral structure—any blob that rotates like a galaxy, with different rotation periods at different distances, will make a spiral pattern. The problem is to find out how a galaxy makes a spiral shape that can persist. Currently, there are three different kinds of answers, and astronomers are not sure yet which is right. It may be that all are right in one or another case and that the spiral structure in even one individual galaxy can have a mixed parentage.

The Density-Wave Theory of Spiral Structure

Perhaps the most elegant explanation of spiral structure in galaxies is known as the density-wave theory. After various related theoretical ideas had been developed in the 1940s by the Swedish astronomer Bertil Lindblad, the density-wave theory was fully worked out and applied successfully to galaxies in the 1960s by C. C. Lin and his students at the Massachusetts Institute of Technology. They showed, through mathematical analysis of the stability of a flat disk of stars, that an irregularity in the initial distribution can stabilize and gradually form into a two-armed pattern of spiral density wavefronts that rotates much more slowly than the stars themselves. By replacing material arms with slowly propagating density wavefronts, these scientists were able to dispense with the winding dilemma.

As stars move through the crest of a spiral density wave, they slow down for a while, enhancing the stellar density along the spiral wavefront by a few percent. As in a moving traffic jam, the delayed and crowded stars eventually move on, to be replaced by new stars "upstream." The spiral-shaped traffic jam itself slowly moves around the galaxy with a fixed pattern speed and so never winds up (see Figure 2.3).

Even more interesting effects are predicted for the gaseous component in the disk. As the more responsive gas enters the wavecrest, the sudden change in velocity induces a supersonic shock wave that compresses the gas manyfold. Dense new clouds form in the shocked gas, and preexisting clouds collide with each other. These dissipative processes among the clouds can trigger star formation, and this explains why, in many spiral galaxies, one finds concentrations of massive gas clouds and newly formed stars in the arms.

The density-wave theory neatly explains the coherent spiral structure observed in grand design galaxies, such as the Messier objects M81 and M74 (see Plate 4). The theory is less successful at explaining the more flocculent kind of spiral galaxy, such as M33, where the spiral arms are typically fragmented, vague, and indistinct.

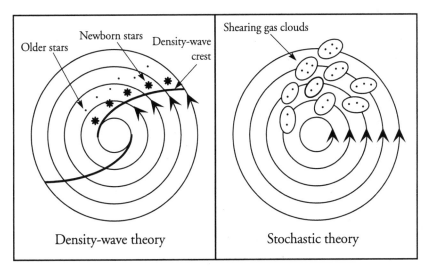

FIGURE 2.3 Spiral density waves versus stochastic self-propagating star formation as modes of generating and maintaining spiral structure. The spiral density wave (left) acts in a coherent manner, compressing all matter that overtakes it. The gas is especially responsive, bunching up along the wavefront and spawning caches of newborn stars downstream. The density wave, being nonmaterial, does not shear and so avoids the winding dilemma. The stochastic mode (right) relies on the shearing of gas clouds in the disk, followed by self-propagating star formation within the clouds, which produces a fragmentary spiral pattern. Each spiral fragment briefly blazes forth with bright young stars and then fades away, thereby avoiding the winding dilemma.

The Stochastic Theory of Spiral Structure

The theory that works best for the less "grandly" designed spiral galaxies builds on the rather simple distortions in any structure that naturally result from a galaxy's rotation. Instead of a persistent set of spiral arms, this idea predicts the continuous birth and death of spiral segments.

Many pioneers in the field realized that this method might work; they merely needed to find a way of regenerating the arms. By 1965 a computer movie was made that showed the entire process in action, using the Andromeda galaxy (M31) as a model and adopting a random (stochastic) pattern for the emergence of star-forming regions.

As these regions are born, they show up as bright patches of activity. Differential rotation then draws them out into long, narrow, spiral-shaped segments. In 100 million years or so, the gas concentrations in the segments are converted into stars, the brightest stars evolve and die, and the spiral seg-

ments fade into obscurity—to be replaced by newly sheared spiral segments of starbirth activity. Through this stochastic process of spiral segment
generation, dissolution, and regeneration, the winding dilemma is handily
avoided (see Figure 2.3).

The result is not a perfect two-armed spiral pattern, of course, but rather a
collection of spiral segments that cover a galaxy and that give it a spiral-like
structure, but with arms that cannot be traced around the center for more
than a few tens of degrees. The spiral features produced by the computer
movie actually resemble those seen in many spiral galaxies, and so it is likely
that a stochastic process of some sort dominates in these more ragged systems (see Plate 4).

The stochastic mode of spiral arm generation also helps to explain certain
kinds of star-forming regions, where spatial sequences of evolving activity are
evident. At the "front" of such regions, there is a giant molecular cloud about
to condense into a star cluster; behind it is a gas cloud illuminated and
cleared out somewhat by the presence of stars just formed; and behind that is
a fully exposed star cluster which is aging and slowly dispersing. This sequence of regions is roughly linear and will be pulled out into a spiral arm
segment by differential rotation. The result is a spiral galaxy made up of disjointed spiral arm segments displaying sequential star-forming and star-
evolving activity.

The stochastic theory seems able to explain just those galaxies that fail to
be accounted for by the density-wave model. Thus we may need no more
ideas, just the patience to obtain the detailed measurements necessary to
check the properties of spiral arms against the various versions of each
theory.

Alternate Theories of Spiral Structure

There is, however, another possibility. Any disturbance of the disk can lead
to a bunching up of the gas in a way that will show up as spiral arms or spiral
segments. The disturbance might come either from outside the galaxy or
from inside the galaxy's nucleus. Of the former possibility, there is the chance
that intergalactic gas might stream into the galaxy, forming arms as it does
so. This idea is not very attractive because the gas would preferentially come
in at the poles, where there is not already gas to collide with, and we know of
very few cases of spiral arms that are not in the plane of the disk.

A more attractive outside agent would be the tidal action of other galaxies
during close encounters. The tides raised by a near collision will involve stars

and gas and can disturb the galaxy's shape enough to lead to an irregular pattern that will rotate into a spiral shape. This is a nice idea, but it has the disadvantage of requiring a close encounter with another galaxy. Typically, the distances between galaxies are too large for the mechanism to explain the observed spiral structure in most disk galaxies.

Statistics aside, several nearby examples of closely interacting galaxies demonstrate the efficacy of spiral arm generation via tidal action (see Chapter 10). Moreover, recent measurements of star-formation rates show that galaxies very near each other seem to have an abnormally high level of star formation going on, especially in their nuclei. Perhaps it will turn out that tidal effects are more easily triggered than we now think.

There is no compelling evidence to suggest that spiral arms might result from action in the nuclei of galaxies, but enough things go on in those mysterious and violent places to make the idea come up. Radio galaxies and quasars (see Chapter 11) all involve highly energetic events in galactic nuclei, many of which eject immense streams of gas well beyond the visible galaxy. Perhaps this kind of activity can somehow lead to spiral arms; at present, though, this idea is vague and uninformed by any reasonable physical model.

Bars

Approximately one-third of the bright spiral galaxies exhibit central bars. These remarkable structures are thought to be more directly related to the flat disk than to the spheroidal bulge component in a spiral galaxy. Typically, the bar involves a concentration of stars that crosses the nucleus, extends symmetrically on each side, and connects with the outer spiral arms (see Figure 1.4 and Plates 1–3). Bars that occur in S0 or Sa galaxies are smooth and made up entirely of stars, while those in Sb, Sc, or Irr galaxies often have lots of gas and dust in them as well.

Measurements of the stellar kinematics in bars indicate that the bars tend to rotate like solid bodies. In some ways the bars behave like standing density waves in the disk—rotating at a fixed pattern speed but consisting of different stars over time. The existence of bar asymmetries themselves is not so very surprising to astronomers who study the dynamics of galaxies. Numerical models predict that instabilities in the disk of a rotating galaxy frequently show up in the form of a bar that resembles those which have been observed.

Arguments still abound concerning the motions of the gas in bars; some evidence shows the gas flowing outward along the bar and other data indicate that it flows inward. Perhaps both occur in what is known as streaming motions. Several theorists have invoked these sorts of motions as ways of bringing fresh material into the nuclei, thus explaining the nuclear activity that is observed in some barred spirals. Recent surveys, however, seem to indicate that barred galaxies have no advantage in hosting nuclear activity.

On firmer footing is the relation between bars and spiral structure. Some of the most prominent spiral structure is observed in barred galaxies. Typically, the spiral arms begin at the tips of the bar and wind around for up to a full turn. Density-wave theorists think that bar asymmetries tidally drive spiral density waves in the disk. Numerical simulations of barred galaxies yield similar results, with the bars generating and maintaining spiral structure over billions of years.

Rings

In addition to bars and spirals, disk galaxies sometimes sport inner and outer rings. Approximately 50 percent of all spiral galaxies play host to inner rings or "pseudo-rings" comprised of tightly wound spiral arms. Many of these stellar and gaseous structures are ablaze with newborn massive stars. The close association between rings and starburst activity indicates dynamics conducive to the formation of giant molecular clouds and of powerful stars therein (see Plate 5). The much rarer and fainter outer rings are only evident in deep astronomical images, where they sometimes show narrower "waists" oriented parallel or perpendicular to whatever central bar may exist.

Examination of rings and their location in disk galaxies reveals some intriguing preferences. Inner rings are often found encircling central bars, or are well inside the bars with radii somewhat less than the semi-minor axes of the bars. They typically measure a few thousand light-years in radius. Nuclear rings are much smaller, with radii of only a few hundred light-years. The ratios of radii between the nuclear, inner, and outer rings are also rather fixed. Here, the density-wave theory seems to best explain what is seen.

Given a spiral density wave with a fixed pattern speed, there will be certain radii where the orbital motions of the stars and gas resonate with the motion of the wave. Like a child being pushed on a swing, the stars and gas at these special radii are periodically and resonantly forced by the wave into ever greater radial excursions. The results are resonantly driven pileups—sometimes near the resonant radii, sometimes in between two nearby resonances.

The presence of a bar provides the most convenient way to generate a spiral density wave and so induce ringlike structures, but it is not absolutely necessary. Any periodic perturber can do the trick. For example, in a similar process, the rings of Saturn are thought to be orchestrated by orbital resonances between Saturn's gravitating moons and the icy chunks constituting the planet's ring system.

How Spiral Galaxies Evolve

Before leaving the spiral galaxies, it is worth pondering how all their marvelous features have evolved over time. If the bulges were the first to form, the disks would have been next—as infalling gas collided and dissipated at the rotational midplane. Instabilities in the disk inexorably led to the formation of central bars. Once established, bars (or companion galaxies) could generate the density waves necessary to orchestrate spiral and resonant-ring structures in the disks.

It is possible that the resonant generation of nuclear rings has fostered the growth of some bulges, which, in turn, has led to the dissolution of some central bars. If bars turn out to be effective conduits of inflowing matter, they could also have helped to grow central bulges and so been the instrument their own undoing. Meanwhile, the mining of energy from the disks to power the spiral density waves may have induced slow radial migrations of stars and gas inward—turning Sc-type galaxies into more centrally concentrated Sb- and Sa-type galaxies. So, it seems that nothing is fixed. Bulges, bars, spirals, and rings may have all undergone significant changes over the past 10 billion years—and may undergo further transformations in the future.

Order among the Irregulars

Irregular galaxies are not completely irregular in their characteristics. They have a few features in common that hint at the reasons for their apparently chaotic shapes (see Plate 3). They are all rich in gas, and almost all of them have lots of young hot stars and clouds of glowing, ionized gas, which are often exceptionally large and brilliant. None has a central bulge or any real nucleus. The light of irregular galaxies, on the average, decreases in intensity outward from the center according to the same exponential law that applies in spiral galaxies. Many irregular galaxies have barlike structures in their central areas, the Large Magellanic Cloud being a particularly good example (see Plate 14 and Chapter 7).

An important hint about how irregular galaxies form comes from a comparison of their total luminosities with those of spiral galaxies. The irregulars are almost all much fainter than even the faintest spiral galaxy. The spiral M33, which represents about the lower limit to the luminosity range for spirals, is still brighter than the Large Magellanic Cloud, which is among the brightest irregulars known. The lack of spiral arms in irregular galaxies, therefore, seems to be related to their low luminosities and corresponding smallness.

The lack of spiral structure also relates to the amount of angular momentum and the amount of turbulent motions within these less massive systems. The disks of irregular galaxies are relatively thicker than those of spirals, which suggests rotational motions not much greater than the turbulent motions. With rotation speeds of 50 kilometers/second or less, these disks are probably too slow to generate spiral arms. If the rotation were even slower, however, the galaxy would not have collapsed to a plane at all, thick or thin, and a low-mass dwarf elliptical galaxy might have formed.

We are not really sure of the relationship between dwarf elliptical galaxies and dwarf irregulars. The traditional view is that the ellipticals contain only ancient stars (10 or more billion years old), while the irregulars have both old and young stars. But there is some evidence to suggest that some dwarf ellipticals—the Carina dwarf, for example—were still actively forming stars only 2 or 3 billion years ago, and during those episodes they may have looked like dwarf irregulars. This is an important issue, because the dynamical explanations of the differences between the two will have to be discarded if it is found that they freely change from one to the other and back again (see Chapter 8).

There is still much to learn about the form and function of galaxies, although we are making progress. We can do more than just describe the differences; we can provide cogent testable explanations for many of them. Meanwhile, the number of unsolved problems—especially regarding the long-term evolution of galaxies—is substantial enough to challenge astronomers for many years to come.

3

GALACTIC ANATOMY

Galaxies are made of many things. In visible-light images, we typically find stars and star clusters, glowing clouds of gas and obscuring lanes of dust—the balance strongly depending on galaxy type. Invisible except to radio telescopes are atomic and molecular gas clouds, along with energetic plasmas consisting of charged high-speed particles. Invisible to every means of detection, so far, are the unknown objects that make up, perhaps, the bulk of a galaxy's mass—those dark yet ponderous forms of matter that are thought to pervade the halos of galaxies (see Chapter 4). We know the most, of course, about the directly detectable things, and they are capable of telling us a great deal about the lives of their host galaxies.

Star Types

The starlight from galaxies is a composite affair, each star contributing its unique spectral energy distribution to the radiant mix. We perceive this variety in terms of the star's color. For example, our naked eyes can see that Betelgeuse in the left shoulder of Orion the hunter has a pale orange tint, Rigel in Orion's right knee has a blue-white hue, Sirius in Orion's companion dog Canis Major is a brilliant white, Procyon in neighboring Canis Minor appears yellow-white, while our own Sun has a distinctly yellowish cast.

Astronomers quantify the color of stars by measuring the light output through different filters and then comparing the relative radiant fluxes in the form of a flux ratio or difference of magnitudes. For example, the $(B - V)$ color of a star corresponds to the difference of magnitudes as measured through a blue (B-band) and yellow (V-band) filter. According to the time-honored system of magnitudes, the fainter the blue flux is relative to the yellow flux, the larger the blue magnitude is, and so the greater the value is of the $(B - V)$ color. By this measure, bluish Rigel has a $(B - V)$ color of -0.02 magnitudes, our yellowish Sun has $(B - V) = +0.65$ magnitudes, and ruddy Betelgeuse has $(B - V) = +2.1$ magnitudes.

It turns out that the spectral energy distributions (SEDs) of stars closely approximate those of hot opaque bodies, more specifically known as black bodies. The ideal black body is a perfect radiator, with no light reflecting and an SED that depends solely on the body's temperature. As the temperature increases, the SED becomes skewed toward higher-energy photons and correspondingly shorter (bluer) wavelengths (see Figure 3.1). Consequently, the measured color of a black body provides a quantitative means of determining the body's surface temperature. Similar color-temperature relations hold for quasi-black bodies, such as the pokers in a blacksmith's furnace, as well as the surfaces of stars. These hot bodies radiate according to the same law, and so can be diagnosed via the same color measurements.

Another way to determine a star's surface temperature is to study its detailed spectrum. This sort of spectral analysis was first carried out in the late 1800s shortly after the invention of the spectroscope. Under the direction of Edward C. Pickering at the Harvard College Observatory, Annie Jump Cannon, Willamina Fleming, Antonia de P. P. Maury, and other pioneering women astronomers classified more than 200,000 stars according to the relative strengths of various absorption lines in their spectra. Originally, the spectral types were ordered alphabetically according to the strength of their hydrogen absorption lines, where stars with the strongest hydrogen absorption lines were of type A. As more was learned about the colors of stars, the sequence was reordered according to increasing redness. Subsequently, absorption lines of calcium, magnesium, iron, sodium, and titanium oxide were included as part of the classification criteria. Eliminating redundancies resulted in the abbreviated sequence, from the bluest to the reddest, of O, B, A, F, G, K, and M spectral types that astronomers still use today. A handy way to remember the spectral sequence is with the mnemonic "**Oh Be A F**ine **G**uy (or Girl), **K**iss **M**e" (see Figure 3.2).

Each spectral type was further divided into numerical subtypes ranging from 0 to 9, the higher number indicating a "lower-energy" spectrum and a redder color overall within the type. According to this system, bluish Rigel

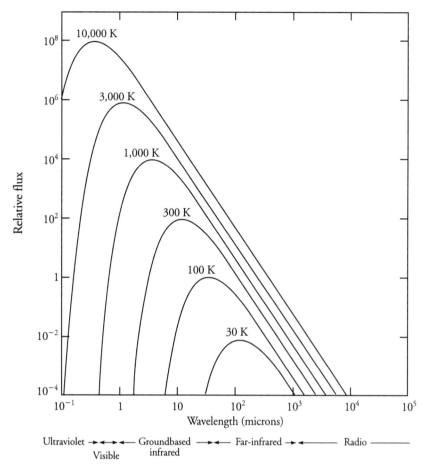

FIGURE 3.1 Spectral energy distributions (SEDs) of black bodies at different temperatures as a function of wavelength. Here, the wavelength is in units of micrometers (10^{-6} meters), also known as microns. As the temperature increases, the wavelength of peak emission shortens, while the luminosity per unit area (surface flux) greatly increases. The temperatures are given in units of Kelvins (K), which correspond to degrees Celsius above absolute zero, such that a temperature in Kelvins is equal to the temperature in Celsius plus 273 degrees.

has a B8 classification, white Sirius an A0, yellow-white Procyon an F5, the yellow Sun a G2, and orange-red Betelgeuse an M2.

Thanks to Cecilia Payne-Gaposhkin, one of the greatest astrophysicists of the twentieth century, we now understand that this sequence of spectral types provides an accurate way to diagnose the surface temperatures of stars. The hottest stars, with temperatures of 30,000 to 50,000 Kelvins (K) (54,523°F to 90,523°F) are O-type stars. Their surfaces are too hot for hy-

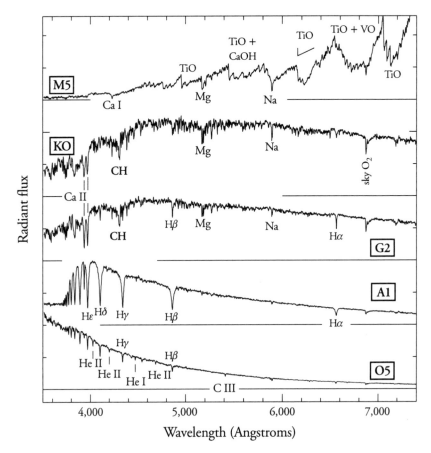

FIGURE 3.2 A sequence of stellar spectra in order of decreasing temperature (from bottom to top) and corresponding spectral type (from O to M). These spectra show only the visible band, with violet wavelengths on the left and red wavelengths to the right. The wavelengths are in the commonly used units of Angstroms (10^{-10} meters); 5,000 Angstroms equal 0.5 microns (see Figure 3.1). The various H-lines (such as Hβ) refer to absorptions produced by atoms of hydrogen—the most abundant of all elements in the Universe.

drogen to retain its single electron, and so they do not show hydrogen absorption lines. But the high temperatures of O stars are amenable to absorptions by the more tightly bound electrons in neutral and singly ionized helium.

Then come the B-type stars, with temperatures of 11,000 K to 25,000 K, where helium absorption lines are weaker but hydrogen lines are becoming more prominent; A-type stars (8,000 K–10,000 K) with the strongest hydrogen absorption lines; F-type stars (6,500 K–7,500 K) with absorption lines from singly ionized calcium; G-type stars like the Sun (5,500 K–6,000 K) with strong neutral calcium, magnesium, and iron absorption lines; K-type stars (4,000 K–5,000 K) with additional absorption bands from cyanogen (CN) molecules; and finally the M-type stars, which have temperatures of only 2,500 K–3,500 K (5,023°F to 6,823°F) at their surfaces and so are sufficiently placid to host loosely bound titanium oxide (TiO) molecules and their corresponding absorption bands.

Cecilia Payne-Gaposhkin included these results in her pioneering Ph.D. dissertation, whose publication in 1925 as *Stellar Atmospheres* became the first monograph of the Harvard College Observatory. As a result of this work she became the first person at Harvard to receive a Ph.D. in astronomy. By providing a physical basis for the spectral types, her work showed that stars are nearly all made up of the same ingredients. Hydrogen clearly dominates, its mass fraction ranging between 70 and 75 percent. Helium is next, amounting to 23–28 percent by mass. The remaining few percent includes all of the heavier elements that are essential to our lives here on Earth.

Our subsequent understanding of the stellar spectral sequence represents one of the major triumphs of twentieth-century astronomy. We now know that the principal differences among the observed stellar types result from the different masses, ages, and chemical compositions that are involved. For a typical assemblage of stars, such as those around the Sun, the key distinguishing difference is just mass. Most stars are found in a stable, relatively unchanging stage of life (known as the main sequence). Most nearby stars also have very nearly the same composition. Mass determines, then, most of the differences we see: high-mass main-sequence stars are hot, very luminous, and blue in color, while low-mass stars are cool, of low luminosity, and red in color. Main-sequence stars of intermediate mass, such as the Sun, are intermediate in their characteristics—fairly cool, of average brightness, and yellow in color.

Stellar Lives

A star spends about 90 percent of its life in its main-sequence phase. During this time, hydrogen is fused into helium in the star's thermonuclear core. The resulting steady release of energy (initially in the form of gamma-ray photons) slowly migrates outward. As the photons scatter off atoms in a sort of drunkard's walk, new photons are generated, the energy per photon decreasing in step with the increasing number of atoms getting excited and new photons being spawned. A million or so years after being generated, the thermonuclear gamma radiation has heated the entire star, and in the process, has been down-converted into a flood of lower-energy ultraviolet, visible, and infrared photons. Finally, the flux of radiant energy makes it to the layer of last scattering, which we see as the visible surface, or photosphere, of the star.

Throughout the star's main-sequence phase, the temperature, color, luminosity, and other observable surface properties remain virtually unchanged. Before reaching this stable state, the star is not really a star: it is a protostar, whose core has yet to begin fusing hydrogen, and whose surface is much redder and brighter than what it will soon become. Most protostars are very rarely seen, however, because the protostellar stage takes up only a tiny fraction (0.1 percent) of a star's total lifetime. We are far more likely to notice a star when it is in its bright post-main-sequence giant stages, which are also short but not nearly as brief as the protostellar phase—amounting to roughly 10 percent of the star's total lifetime.

At the end of the main-sequence stage a typical star becomes a red giant (very massive stars become supergiants); its volume enlarges, its surface temperature decreases, its color reddens, and its luminosity increases dramatically. During this period even low-mass stars become brazenly conspicuous, and so we are most apt to detect them in distant environments, such as other galaxies.

When we look at the stars in a nearby galaxy, we see typically the most luminous stars—the bright blue, massive main-sequence stars and the red evolved giants of lower mass. The relative balance of these respective populations tells us a lot about the star-forming history in the galaxy, because the life expectancy of a star depends very much on the star's mass.

The most massive stars live only a short time, a few million years, whereas the intermediate-mass Sun has been a main-sequence star for about 5 billion years and has another 5 billion to go before it turns into a red giant. For this reason, we will not see bright, blue, massive stars in a galaxy that has not been forming stars recently. For an elliptical galaxy, where most of star formation occurred billions of years ago, we will detect only the long-lived intermediate-mass stars that have just recently become red giants.

Star Stuff

Although chemical abundances in most stars are very nearly the same, there are small differences that reveal much about the stars' respective pedigrees. They do not affect the star's overall luminosity or apparent color very much, but the differences can be detected in the spectra and quantitative colors of the stars. One such effect is observed in the far-violet (U-band) part of the spectrum, where the number of iron absorption lines is sufficient to effectively "blanket" the emerging far-violet light, and so redden the star's ($U -$ B) color. This photometric index of spectral-line blanketing provides an especially useful way to estimate the abundance of iron and other heavy elements in stellar clusters, where detailed spectroscopic observations and analyses of the many stars would be too time consuming.

As Cecilia Payne-Gaposhkin found, the vast bulk of almost all stars is comprised of hydrogen and helium. For example, most stars in our area of the Galaxy have only about 2 percent of their mass in the form of other heavier elements. In a few stars, an even smaller percentage of their mass is made up of these heavier elements, down to values as small as 0.01 percent. Because the metallic elements are most prominent in stellar spectra, astronomers tend to call all of the heavier elements "metals." Stars like the Sun are considered metal-rich, while those depleted in heavy elements are regarded as metal-poor. In the following section, we will see that the distribution of metal-rich and metal-poor stars in a galaxy provides important but sometimes confusing clues to the evolution that has occurred over the galaxy's lifetime.

Stellar Populations

Walter Baade was a German émigré to the United States during the time of Hitler's rise to power. Having settled in California, he worked at the Mount Wilson Observatory near Los Angeles—then the most important of the world's observatories. During World War II, many of his American colleagues were called upon to aid the U.S. war effort. Baade was classified as an enemy alien, and so was disqualified from participating in war work. Consequently, he had a virtual monopoly of observing time on the 100-inch Hooker telescope atop Mount Wilson. The enforced blackouts of Los Angeles further aided him in his efforts to obtain deep and detailed photographs of nearby galaxies.

Concentrating on the nearest giant spiral galaxy M31 and on its dwarf companions (see Plate 16), he was able to decompose the starlight into individual stars and so make comparisons with stars and star clusters in our Milky Way galaxy. By the 1950s, he had developed the two-population scheme that continues to underlie our concepts of stellar and galactic evolution.

Baade showed that the two populations included very different kinds of objects and occupied different parts of galaxies. Population I included very hot, very luminous stars (designated O and B stars) along with some cooler stars, open star clusters, young stellar associations, atomic hydrogen gas, and dark lanes of dust. The stars constituting Population I were later found to be fairly young and seemed to contain abundances of heavy elements similar to those measured in the Sun, where about 70 percent (by mass) is hydrogen, 28 percent is helium, and 2 percent is everything else.

As originally defined by Baade, Population II stars included red giants and other faint main-sequence stars that were typically associated with these giants; bright blue stars, gas, and dust were excluded. Groupings of Population II stars were restricted to globular clusters in the halos of galaxies. The stars were later determined to be very old (on the order of 10 billion years old or more) and deficient in heavy elements—typically 10 to 100 times poorer in such elements than the Sun.

Population I objects were found almost exclusively in the disks of spiral galaxies and in irregular galaxies like the Magellanic Clouds. By contrast, Population II objects inhabited the central bulges of spiral galaxies and their tenuous outer halos. Elliptical galaxies were considered to be pure Population II objects.

Baade's concept of the two populations was simple and astonishingly successful. It helped to unravel the complex interplay between age, dynamics, and element production in galaxies, and catalyzed our understanding of stellar evolution. But, like many scientific breakthroughs, it was eventually found to be an oversimplification. The grand scheme of two types of stars did not take into account intermediate objects of one sort or another and completely omitted all kinds of exceptions. At first, astronomers were tempted to elaborate on the scheme, defining various subtypes (Population IIa, Population I.5, and so on), but eventually the futility of doing so became apparent.

As our understanding of galactic properties increased, and as modern equipment improved, the need for a simplified scheme vanished. Baade's scheme had served its purpose admirably, but now astronomers can use more quantitative measures of the characteristics of a group of stars. Now an actual age and elemental composition can be assigned to a star cluster or a portion of a galaxy. Over the years, Baade's Population I and II were used less and less, until today astronomers hardly ever mention them.

One of the reasons Baade's original scheme broke down was the striking correlation between age and chemical composition that was evident in the Milky Way. Population I was young and rich in heavy elements. Population II was old and poor in heavy elements. This correlation was beautifully explained in 1959, when it was realized that heavy-element production in stars causes a gradual change in the chemical abundances of subsequent stellar populations. While the early-forming stars are almost pure hydrogen and helium, those that formed later on, especially those that are forming now, are made up of enriched material that was produced in the interiors of past generations of stars. Our Sun and Solar System, only 4.6 billion years old, has heavy elements in it that were forged in the thermonuclear cores of previously blazing stars. In our Galaxy, this enrichment process is embodied in the Population I and II stars, with the Population II stars in the halo being ancient and metal-poor, and the Population I stars in the disk being much younger and metal-rich.

But in the 1960s the Large and Small Magellanic Clouds began to present a problem to astronomers. First, these two companion galaxies to the Milky Way contained many globular clusters of low metallicity that were young rather than old. Furthermore, many young stars were found elsewhere in the Clouds, which, when subject to detailed spectroscopic analysis, also turned out to be poor in heavy elements. Because age and metal abundance were no longer correlated, the Population I and II classifications could not be applied to these galaxies.

In recent years, astronomers have found that many of the supposedly pure Population II elliptical galaxies are surprisingly rich in heavy elements, as are the central bulges of spiral galaxies. Even in our Galaxy there is a breakdown of the scheme; at the very outer parts of the Galactic disk the heavy-element abundance in young stars is unexpectedly low, whereas the chemical abundances are much higher in the ancient stars that inhabit the central bulge component and the innermost globular clusters. Once again, the abundances, locations, and ages are no longer in lock step.

Rather than relying on Baade's outmoded categories, we now use specific measures (ages, abundances, dynamics, location) to characterize a population. For example, we often find that the metal abundance correlates with the overall luminosity (and hence mass) of the system, while the age of a population correlates with its deviation from co-planar circular orbits. Some of the ways by which astronomers measure these salient properties are discussed next.

Color-Magnitude Diagrams

A handy way to find out what kinds of stars exist in a region of space is to make a plot of their individual colors and magnitudes (see Figure 3.3). The result is called a color-magnitude diagram (CMD). On it you can plot not only the stars that you observe, but also the expected distribution of stars of different masses, ages, and compositions. By plotting the observational data

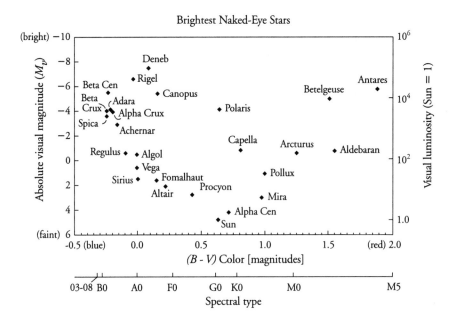

Brightest Naked-Eye Stars

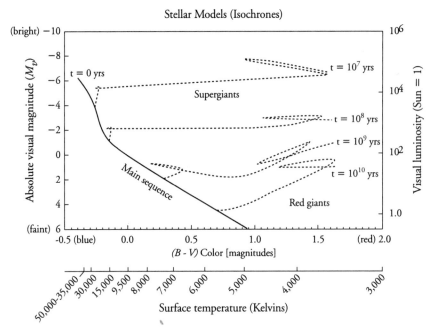

Stellar Models (Isochrones)

FIGURE 3.3 (facing page)   Color-magnitude diagrams. Top: Color-magnitude diagram for well-known stars in the naked-eye sky. The vertical axes denote their absolute visual magnitudes and corresponding luminosities (from faint at the bottom to bright at the top), while the horizontal axes shows their $(B - V)$ color indices (from blue at the left to red at the right) and their corresponding spectral types. The stars on the diagonal band (from upper left to lower right) of this diagram are on the main sequence, while most of the stars to the right are evolved giants and supergiants. Bottom: The same diagram with theoretical isochrones, where each isochrone represents a stellar population of a particular age. According to these models, the supergiants Deneb, Betelgeuse, and Antares are much younger than the relatively dimmer red giants Arcturus, Aldebaran, and Mira.

with the theoretical models on the same CMD, astronomers can infer the basic properties of the observed stellar populations.

For stellar systems beyond the Milky Way galaxy, only the bright stars can be seen, but these are still sufficient to yield vital information about the stellar populations. In a galaxy like M31, for example (see Chapter 9), we can see the young, blue, massive stars on the bright part of the main sequence as well as the red, evolved giant stars, and so can figure out the relative numbers of young and old stars. This effort has been helped enormously by the high-resolution capabilities of the Hubble Space Telescope as well as the latest groundbased telescopes, whose actively controlled optical elements can compensate for the distortions produced by the Earth's atmosphere. Color-magnitude diagrams featuring thousands of stars are now commonplace. In Part II we will pursue this CMD technique in diagnosing the star-forming histories of individual galaxies.

Stellar Luminosity Functions

Galaxies close enough for color-magnitude diagrams to be obtained are also resolvable enough so that the numbers of stars of various brightnesses can be counted. This enumerating process produces what is called the stellar luminosity function. Within 100 light-years of the Sun, a complete census of the stars can be made down to those with very faint luminosities. The resulting stellar luminosity function is shown in Figure 3.4, where the number of stars within a given volume of space is plotted as a function of absolute magnitude. Stars of subsolar luminosity are seen to greatly outnumber those like the Sun, which in turn are far more numerous than the most luminous stars.

The stellar luminosity function in the Solar Neighborhood (within 100 light-years of the Sun) is statistically complete down to much lower luminosities than can be detected in more distant stellar populations. As such, it pro-

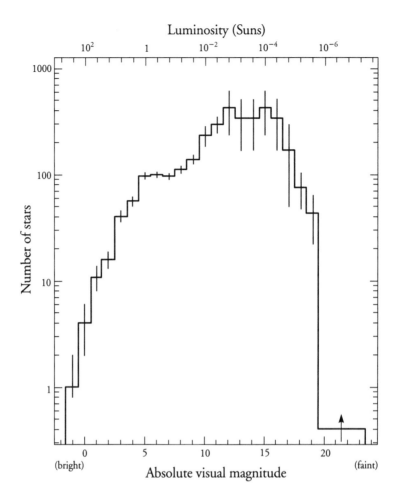

FIGURE 3.4 The stellar luminosity function for stars in the Solar Neighborhood. The vertical axis denotes the logarithmically scaled number of stars as a function of absolute magnitude, itself a logarithmic measure of luminosity. The horizontal axis marks the absolute magnitude (M_v), with the corresponding luminosity decreasing from left to right. For reference, the Sun has an absolute visual magnitude of 4.85 mags. Note the dominance of faint (and hence low-mass) stars. The bump at $M = 5.5$ mags is likely due to enhanced star formation having occurred for stars with luminosities of about half that of the Sun, and masses of about 0.8 the solar mass. These stars have total burning lifetimes of roughly 12 billion years—coinciding with the estimated age of our Galaxy's disk.

vides an essential benchmark for evaluating the limited luminosity functions that are obtained in these more distant systems. The Solar Neighborhood's luminosity function also provides vital clues to the history of star formation that has played out in our part of the Galaxy. For example, the observed bump at an absolute visual (*V*-band) magnitude of $M_V \cong 5.5$ mag probably indicates an overabundance of stars with characteristic ages of 11–14 billion years, and hence a burst of star formation having occurred 11–14 billion years ago.

Stellar luminosity functions are now available for many galaxies in the Local Group and for a few more distant ones. In the Large and Small Magellanic Clouds (LMC and SMC), the luminosity functions show a relative overabundance of high-luminosity stars compared to what is observed in the Solar Neighborhood. This strong showing of young, super-luminous stars and star-forming gas in the Magellanic Clouds was previously thought to be a general feature of irregular galaxies. But we now recognize that some dwarf irregular galaxies are dominated by intermediate-age and older stars, with even fewer bright blue stars than are seen in the Solar Neighborhood. Star-forming irregular galaxies like the LMC and SMC may turn out to be more the exception than the rule (see Chapters 7 and 8).

The stellar luminosity functions of elliptical galaxies, such as the Sculptor dwarf galaxy, look much like those of the ancient globular star clusters. Here, a low-luminosity peak in the numbers of stars suggests that a corresponding peak in the star-formation rate occurred 12—15 billion years ago. More recent work has shown that Sculptor and other nearby dwarf elliptical galaxies have populations of intermediate as well as ancient stars, indicating more complex star-forming histories (see Chapter 8).

Spectral Types of Galaxies

The spectra of galaxies also provide important clues to the types and numbers of stars that are present, and can yield some indications of the chemical compositions in these stars. The spectra are not simple to interpret, however, because they are composite, and it is not always easy to unravel the blended effects of the myriad stars contributing to each spectrum. Because of the importance of spectroscopy in diagnosing the stellar content in galaxies, considerable effort has gone into obtaining highly accurate spectra. When this information is combined with the distributions of stars, gas, and their corresponding motions in a galaxy, a fairly complete model of the galaxy emerges.

Most recent studies of galaxy spectra use an electronic detector, which accurately measures the brightness in the spectrum at a large number of differ-

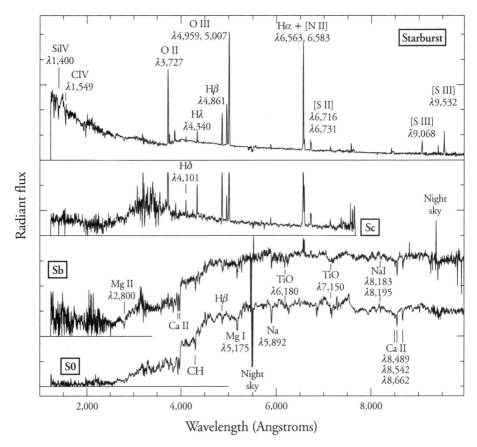

FIGURE 3.5 Ultraviolet-optical spectra of representative spiral galaxies, from a starburst (at the top) to an S0 type (at the bottom). Note the change in overall emission from blue-dominant in the starburst to red-dominant in the S0. The starburst spectrum is dominated by emission lines, much like those seen in individual H II regions, that betray the presence of newborn massive stars of high surface temperature. At intermediate types, the various atomic absorption-line and molecular absorption–band features indicate a composite population of hot main-sequence and giant stars, which are typically less than a few hundred million years old, along with cool red-giant stars, which tend to have ages of several billion years.

ent wavelengths. The observational data are then modeled by computer according to various assumed mixes of star types and chemical compositions. With a large enough library of individual stellar spectra spanning a sufficient range of ages and chemical abundances, astronomers can derive a reasonable fit to the types of stars constituting a particular galaxy.

For example, in Sb-type galaxies such as M31, the various spectral-line indices make it quite clear that the stellar population is composite (see Figure 3.5). Hydrogen lines (Hβ and Hδ) appear in the blue part of the spectrum, whereas titanium oxide bands (TiO) are present in the far-red. One has to conclude that there are some stars hotter than the Sun to give the hydrogen lines and some stars much cooler than the Sun to host the titanium oxide molecules. At the same time, there must be stars roughly similar to the Sun, because of the strong magnesium (Mg), calcium (Ca), and sodium (Na) lines. The composite nature of the Sb galaxy spectra immediately tells us that these galaxies have been making stars for a very long time and continue to do so today—albeit at lower rates.

Interstellar Gas and Its Detection

Most of the gas between the stars in a galaxy is in the form of cold molecular hydrogen, called H$_2$, and cool, atomic hydrogen, called H I. Near hot stars (and active galactic nuclei) the hydrogen is ionized and glows visibly, forming what is referred to as an H II region. Of all the stuff in galaxies, the H II regions are the most dazzling. The Roman numeral after the H refers to the amount of ionization (the number of electrons lost by the atom because of interactions with energetic atoms or photons); a nonionized (neutral) gas of atoms is represented by "I," a singly ionized gas by "II," and so forth. There are small amounts of other gases in H II regions—for example, helium, oxygen, sulfur, and nitrogen—which have more electrons and therefore can have higher amounts of ionization. For example, twice-ionized oxygen (O III) is often detected in the spectra of H II regions containing especially hot stars.

In addition to the discrete H II regions, star-forming galaxies like the Milky Way also contain an abundance of diffuse ionized hydrogen that may contribute as much mass as the molecular and atomic gas components. The hottest and most diffuse gas phase is known as coronal gas, having been first discovered in the Sun's outer atmosphere, or corona. This plasma of highly ionized atoms and liberated electrons has characteristic temperatures of several million Kelvins. It has been detected in the halo of our Milky Way galaxy, where the distinct ultraviolet absorption-line signature of five-times ionized oxygen (O VI) has been found in the spectra of distant hot stars and background quasars. Coronal gas is also detectable in emission at X-ray wavelengths. Spaceborne X-ray observatories have mapped this emission from supernova remnants (the shock-heated debris of recently exploded stars) such as the Crab Nebula, from elliptical galaxies that are otherwise virtually bereft of gas, and from the space between densely clustered galaxies.

These sundry phases of molecular, atomic, and ionized gas cohabitate within galaxies, while maintaining their distinctive temperatures, densities, and pressures. To observe these various phases, astronomers use a variety of techniques spanning a vast range of wavelengths. Sometimes indirect methods of detection must be invoked. For example, the molecule of diatomic hydrogen (H_2) is an inefficient radiator—requiring shocks or other energetic processes to stimulate its characteristic emissions at ultraviolet and infrared wavelengths. A brighter and hence better tracer of H_2 is provided by the carbon monoxide (CO) molecule. Despite its much lower abundance (by a factor of 12,500), CO produces strong line emission at millimeter wavelengths.

The brightest CO emission, produced by the transition from the first excited rotational energy level to the groundstate, occurs at a wavelength of 2.6 millimeters and corresponding frequency of 115 gigahertz (GHz) (115×10^9 cycles/second). This emission was first detected in the late 1970s, when microwave technologies were just beginning to emerge. Today, large radio telescopes specialized for the higher tolerances of millimeter-wave observations routinely map the CO emission from molecular gas in both nearby and remote galaxies.

Other interstellar molecules that have been detected at radio wavelengths include the hydroxyl radical (OH), water (H_2O), ammonia (NH_3), hydrogen cyanide (HCN), formaldehyde (H_2CO), methyl alcohol (CH_3OH), and enough ethyl alcohol (CH_3CH_2OH) to produce galaxy-wide hangovers.

Cool atomic hydrogen, H I, is as abundant as H_2 but typically is more widespread—often extending far beyond a galaxy's optically luminous radius. H I is detected by virtue of the 21-centimeter line emission that is produced when its sole electron reverses its spin angular momentum from being co-aligned with the proton's spin to being in an antiparallel spin state of lower energy. In 1944, the Dutch astronomer Hendrik van de Hulst predicted that the H atom's 21-centimeter wavelength (1.42 gigahertz frequency) emission could provide an observable tracer of interstellar hydrogen. The Galactic emission was detected 6 years later by Harold Ewen, a graduate student working under the Harvard physicist Edward Purcell, who used a crude horn antenna placed outside his laboratory window.

During the 1970s and into the 1980s, Dutch astronomers produced the most detailed maps of H I in the Milky Way and beyond. They used the Westerbork array of radio telescopes in Holland, on the site of what was once a Nazi concentration camp. Today, exquisitely detailed maps of the H I in nearby galaxies are being made with the Very Large Array (VLA) of radio telescopes in New Mexico and the Australia Telescope Compact Array (ATCA) in southeastern Australia.

Ionized hydrogen is made visible when free electrons in the ionized gas recombine with their host protons to form neutral hydrogen atoms. The subsequent cascades to lower energy levels produce spectral-line emission at radio, infrared, optical, and ultraviolet wavelengths. This happens in H II regions, where the number of ionizations by ultraviolet photons from hot stars balances the number of recombinations back to the atomic state, and in supernova remnants, where the shock-ionized gas cools and recombines just behind the expanding blast wave. Starburst galaxies contain a multitude of giant H II regions, resulting in spectra that are indistinguishable from those of individual H II regions (see Figure 3.5.).

The ruby-red hydrogen-alpha (Hα) line at a wavelength of 6,563 Angstroms (6,563 $\times$ 10^{-10} meters) is most frequently used to trace ionized hydrogen, because it is both bright and readily detectable with current electronic cameras. It occurs when the hydrogen atom's sole electron falls from the third lowest energy level to the second lowest level. Other common tracers of ionized gas include the optical line emission from [O II] (3,727 Angstroms) in the far-violet, He II (4,686 Angstroms) and Hβ (4,861 Angstroms) in the blue, [O III] (5,007 Angstroms) in the green, [N II] (6,584 Angstroms) and [S II] (6,731 Angstroms) in the red, and [S III] (9,532 Angstroms) in the far-red (see Figure 3.5). Here, the brackets denote "forbidden" (slow and hence low-probability) transitions that nonetheless occur in the sanctuary of interstellar space, where de-exciting collisions are rare. At infrared wavelengths (which are typically measured in units of micrometers (10^{-6} meters), or microns, new instrumentation is enabling astronomers to detect the hydrogen emission lines of Brackett-gamma (Brγ) at 2.16 microns and Brackett-alpha (Brα) at 4.05 microns, both of which involve electron jumps to the fourth lowest energy level, as well as the [Fe II] (1.65 micron) line, and various other lines from ions of sulfur, neon, argon, and carbon.

At radio wavelengths, ionized hydrogen produces a type of continuum emission known as thermal Bremsstrahlung radiation. It occurs when the free electrons career past protons in the gas, losing energy in the process. Maps of this radio emission in the Milky Way have enabled astronomers to identify distant H II regions that otherwise would be obscured at optical wavelengths.

Another form of radio emission from the interstellar medium is produced from high-energy electrons as they spiral around magnetic field lines at speeds close to that of light. This so-called synchrotron radiation was first identified in particle accelerators, where the same highly magnetic and relativistic conditions prevail. It can be distinguished from thermal Bremsstrahlung radiation by its steeper falloff in power with increasing frequency and its high degree of polarization. From maps of radio synchrotron radiation, astronomers have identified supernova remnants in our Milky Way galaxy and evidence of vast explosions from other galaxies (see Chapter 11).

Gas in Spiral and Irregular Galaxies

The ionized, atomic, and molecular gas phases are most abundant in spiral and irregular galaxies, where they constitute up to 40 percent of the total mass. Sa-type galaxies show the faintest optical emission lines, while irregular galaxies have the strongest lines and the highest proportion of H I gas to stars. The H I gas fraction in galaxies increases steadily through the progression from Sa (2 percent) to Sc (10 percent) to irregular (20 percent). The H_2 component roughly doubles these gas percentages. The distribution of the gas also depends on the type of galaxy. For example, H II regions and their underlying caches of hot stars are often narrowly limited to inner rings in Sa and Sb spirals, but are widely arrayed along spiral arms in Sc galaxies, and are randomly situated in irregular galaxies (see Plates 2–5).

Barred SBb and SBc galaxies often host large amounts of ionized gas in their nuclei, whereas only a few unbarred Sb and Sc galaxies do, thus suggesting different sorts of dynamics orchestrating the gas and subsequent star formation. Perhaps the gravitational effects of the bars help to direct starforming gas from the disk into the nuclei.

Generally, the spectra of galaxian H II regions indicate physical conditions similar to those in normal emission nebulae in our Galaxy, with temperatures of about 10,000 K and electron densities of a few hundred per cubic centimeter. The strengths of oxygen and nitrogen emission lines relative to those of hydrogen decrease away from the centers of many spiral galaxies. Apparently, this is due to a decrease in the abundances of nitrogen and oxygen relative to hydrogen, but other effects may be involved. If we are, in fact, seeing an abundance effect, then we are tempted to conclude that spiral galaxies evolve chemically from the inside out. In barred spirals, the radial gradient in these line ratios is less apparent, perhaps because of the radial mixing of gas induced by the gravitating bars.

Among the few galaxies for which complete CO, H I, and H II surveys have been made, there is a great deal of variety. Unlike M31, where the CO emission, neutral atomic hydrogen, and H II regions are all concentrated in the same large ring, many galaxies show conspicuous differences in the spatial distributions of molecular, atomic, and ionized gas. Typically, the atomic hydrogen gas extends farther from the center than the molecular and ionized components.

This is possibly the result of the way in which gas has condensed into stars over the lifetime of the galaxy. The most efficient star formation has probably occurred near the center of the galaxy, with the completeness of condensation into molecular clouds and subsequent stars decreasing outward from the center. Indeed, it is not at all clear what active role—if any—the more widespread neutral atomic hydrogen plays in the star-formation process. Perhaps the neutral hydrogen acts more like a vast reserve that the galaxy can draw upon for future star formation, if and when the conditions are right.

Irregular galaxies show neutral hydrogen spread throughout their visible disks in a fairly chaotic pattern. In the Magellanic Clouds, for example, recent maps show a froth of hydrogen shells and associated voids. The causal relationships between these gaseous structures and their neighboring stellar associations, H II regions, and supernova remnants remain unclear.

Gas in Elliptical Galaxies

Interstellar gas is an inconspicuous component of most elliptical galaxies. By mass, the most important phase is the hot tenuous gas that can be seen in X-rays, delineating the halos of some ellipticals. This coronal gas, constituting as much as a few percent of the galaxy's total mass, is not primordial in origin. Instead, it is thought to represent the accumulated byproducts of evolving and outgassing stars over cosmic time. According to this scenario, the high temperatures result from the orbital motions of the outgassing stars. The gas inherits the high velocities (several hundred kilometers per second) and randomly oriented orbits of its progenitor stars. The result is a chaotic ensemble of particles that thermalizes into the million-degree phase that is observed.

Spectrographic surveys at optical wavelengths indicate line-emitting "warm" gas in the nuclei of many elliptical galaxies. The intensity of the emission lines ranges smoothly from barely detectable to strong and conspicuous. High-resolution spectroscopy of this warm nuclear component often reveals very large shifts in the wavelength of the observed emission. These shifts are interpreted as being due to the Doppler effect (similar to the changing pitch of an ambulance siren as it approaches and then recedes from you). The large observed Doppler shifts in the nuclei of these galaxies indicate rapid gas motions. Some of the best evidence for super-massive black holes is derived from these motions, whereby the black hole provides the gravitation necessary to hold onto the whirling gases.

In dense clusters of galaxies (such as the Perseus and Coma clusters), elliptical galaxies predominate over their spiral counterparts. Virtually no emission lines are present in the clustered elliptical galaxies, however, probably because collisions between densely packed galaxies have swept them clean of gas. Instead, we find the gas outside the galaxies in the form of a tenuous intracluster medium, diffusely glowing in X-rays at inferred temperatures of 10–100 million Kelvins.

Dust and Organic Matter

Grains of microscopic dust are thought to condense out of the cool atmospheres of aging red giant stars. The dust grains readily mix with the interstellar gas, thus providing another observational tracer of the overall interstellar medium. Reliable data on the properties and amounts of dust in galaxies are not easy to obtain, however. The three ways to detect the dust are through its reddening effect on the colors of a galaxy's stellar components (by observing star clusters and individual stars of known intrinsic color), through its obscuring effects, and through the infrared light that the dust itself emits.

For some galaxies, such as M31, color measurements of the globular star clusters indicate that light absorption by dust depends on distance from the galactic center. Although photographs of spiral galaxies typically show local irregularities in the extinction due to dust, the overall distribution of obscuring dust tends to be symmetrically oriented about the center, decreasing outward and rising in the vicinity of spiral arms.

The method of studying dust in a galaxy by counting distant galaxies seen through it has been applied to several nearby objects, including the very loosely structured dwarf elliptical galaxies in the Local Group. From the distribution of background galaxies counted over the areas of these dwarfs, it is obvious that there is very little absorbing material in them, since the distant galaxies show no decrease in number behind them.

Distant galaxies behind the Magellanic Clouds can also be detected, and the resulting counts have provided astronomers with maps of the total amount of obscuring matter, especially in the Small Cloud. At the center of that galaxy, according to these studies, there are about 1.2 magnitudes of obscuration. Thus, only about one-third of the light gets all the way through.

Another way to study dust in galaxies is to examine the dust lanes seen in projection against the amorphous stellar background of the galaxy. For some galaxies, especially those viewed close to edge-on, such lanes are conspicuous components (see Figure 3.6). An examination of the *Carnegie Atlas of Galaxies* or any other photographic survey of galaxies shows that dust is closely associated with the spiral patterns of most spiral galaxies. It is also strongly concentrated to the plane in most cases.

FIGURE 3.6   Dust lanes in the edge-on galaxy NGC 891, silhouetted against its stellar bulge and disk. Our own view of the Milky Way looks very similar to this (see Plate 6).

Consider the dust in the spiral arms of the Whirlpool galaxy, NGC 5194. This Sbc-type galaxy appears to lie slightly in front of its SB0-type galaxy companion, NGC 5195 (see Plate 22). One of the arms crosses half of the image of NGC 5195, and the dust in this arm shows up clearly as an absorption lane crossing the smaller galaxy. Because NGC 5195 is a relatively symmetric smaller object, it is possible to measure the total amount of extinction caused by dust in the superposed spiral arm. The galaxy's color directly behind the dust is anomalously faint and red. As anticipated, the excess reddening is very nearly proportionate to the total amount of absorption. Furthermore, the dust is distributed over a wider area than the luminous part of the spiral arm. The dust is most conspicuous in the inner parts of the spiral arm, a fact that is also observed in other galaxies. The total absorption caused by the dust, approximately 0.4 magnitudes, is of the same order as the thickness of the dust detected in the local spiral arm of our own Milky Way galaxy.

One of the best ways to map the distribution of dust in galaxies is to observe its emission at mid- to far-infrared wavelengths. When irradiated by the ambient starlight, grains of dust achieve temperatures of 10 K to 100 K. The corresponding spectral energy distributions of thermal (black-body) emission from the heated grains are brightest at wavelengths that are inversely proportional to the dust temperature. At temperatures of 10 K, the emission peak is at the far-infrared wavelength of 290 microns (290×10^{-6} meters). And at 100 K, the dust emission peaks at the mid-infrared wavelength of 29 microns.

For reference, humans have temperatures of 37°C (310 K), which corresponds to an SED with a mid-infrared peak at 9.3 microns. Imaging of human bodies at these infrared wavelengths provides important clues to the physiological processes and anomalies taking place just under our skins. By analogy, astronomers use mid- and far-infrared imaging to diagnose interstellar dust, its powering by nearby stars, and its variations within and among galaxies.

In the 1980s, the Infrared Astronomical Satellite (IRAS) surveyed the entire sky at wavelengths of around 12, 25, 60, and 100 microns. The resulting maps of dust in the Milky Way revolutionized our concepts of the Galactic interstellar medium (see Plate 8 and Chapter 6). IRAS also provided our first low-resolution maps of the dust emission from nearby galaxies. In this way astronomers obtained the first hints that the emitting dust shows both a clumped component along the spiral arms that is most closely associated with H II regions and their host molecular clouds, and a more diffuse component of lower dust-grain temperature that traces the atomic H I distribution.

More recently, the Infrared Space Observatory (ISO) has provided much higher-resolution maps at mid-infrared wavelengths. These maps clearly delineate the emitting dust along the spiral arms, thus verifying the role played by spiral-arm dynamics in orchestrating both the solid and gaseous phases of the interstellar medium.

Perhaps even more exciting is the discovery of galaxian mid-infrared band emission from complex molecules known as polycyclic aromatic hydrocarbons (PAHs). This class of organic molecules—based on the benzene (C_6H_6) ring—is known for its stability in the presence of high temperatures and harsh radiation fields. On Earth, PAHs are associated with combustion processes as familiar as the charring of meat on a grill.

The discovery of PAHs in galaxies opens up all sorts of questions regarding interstellar organic chemistry. Are the PAHs "cooked up" in the H II regions, where we find them glowing most profusely? Or are they ubiquitous throughout the interstellar medium, requiring nothing more than gas clouds enriched with the carbonaceous exhaust of red giant stars? And what do the PAHs tell us about the potential for life among the clouds? Such questions are prompting intense research by laboratory physicists and chemists in concert with observational and theoretical astrophysicists.

Nuclear Pyrotechnics

Our census of the galaxian hoard would not be complete without a look at the wild goings-on in the nuclei of some galaxies. Although rare, the visibly active nuclei betray the presence of bizarre denizens that may lurk inside many more galaxies.

Some of the activity is fairly familiar to us—resembling overgrown versions of the H II regions that highlight the disks of spiral and irregular galaxies. These show bright hydrogen line emission at optical wavelengths, strong Bremsstrahlung and synchrotron radiation at radio wavelengths, gobs of intense CO emission at millimeter wavelengths, and colossal outpourings of light at mid- and far-infrared wavelengths. From what we know about nearby H II regions, we can explain this mélange of emissions as coming from dense concentrations of newborn hot stars and associated supernova activity.

Fuel for the "fire" is indicated by the CO emission and the molecular gas clouds that it traces. The resulting caches of hot O-type stars irradiate the surrounding gas with high-energy ultraviolet photons—ionizing the hydrogen, sulfur, and oxygen, and so inducing copious line emission at wavelengths characteristic of these elements. The freed electrons in the 10,000 K gas occasionally veer past naked protons, their loss of kinetic energy converting into thermal Bremsstrahlung emission. Electrons that have been accelerated by supernova blasts to relativistic speeds interact more strongly with the ambient magnetic fields. The resulting loss of energy is seen in the form of synchrotron radiation. Sometimes plumes of synchrotron-emitting electrons can be observed extending thousands of light-years from the central sites of intense star-forming activity.

Then there is the dust. It efficiently absorbs whatever ultraviolet light has not interacted with the gas, heats up to about 100 Kelvins, and reradiates the energy at mid- and far-infrared wavelengths. The infrared luminosities can be staggering, amounting to the luminosity of 10–100 billion Suns—all contained within a region no larger than a few hundred light-years across. Astronomers dub these regions nuclear starbursts. We will take a more detailed look at the starburst phenomenon in Chapter 10.

When it comes to dishing out the most dazzling fireworks across the widest spectrum of electromagnetic energies, active galactic nuclei (AGNs) can't be beat. First noted as unresolved sources of strong radio emissions, AGNs were later found to radiate profusely at infrared, optical, ultraviolet, X-ray, and gamma-ray wavelengths.

Follow-up monitoring of AGNs has shown that they are highly variable. Outbursts seen at radio and X-ray wavelengths are often followed by brightenings of the optical line emission several weeks later. Astronomers have used the observed timescales of the outbursts and subsequent brightenings to delineate the dimensions of the emitting sources. Because light from the outbursting source took only a few weeks to reach and excite the line-emitting clouds, these clouds must be no more than a few light-weeks from the central source. Similar light-travel arguments indicate that the central outbursts come from a region less than a few light-days across. We are faced with upward of a trillion solar luminosities coming from a piece of real estate no bigger than our Solar System!

Such extreme power requirements combined with the broad emission lines and hence high Doppler velocities that have been measured in AGNs strongly argue for super-massive black holes as the "monsters" that underlie the AGN phenomenon. A black hole with a mass of a million or more Suns would have sufficient gravity to simultaneously bind the rapidly moving gas and power the observed emission (through infall of clouds and the subsequent release of gravitational energy). Interactions between the infalling gas and the intense environment surrounding the black hole might then accelerate the vast radio-emitting jets that are seen squirting out of some AGNs.

All types of galaxies can host active nuclei, although most activity is seen in the giant ellipticals and bulge-dominated Sa- and Sb-type giant spirals. Overall, the number of bright galaxies with noticeable AGNs amounts to a mere 1 percent or so. But very low activity may be present in most large galaxies. Our own Milky Way galaxy would not be counted as one with an AGN. Nevertheless, our proximity to the Milky Way's nucleus allows us to see that it is a very special place, with lots of interesting structure and energetics at radio wavelengths (see Chapter 6).

Galaxies with barely noticeable AGNs are known as LINER galaxies (short for low-ionization nuclear emission region). Perhaps our Galaxy's nucleus would qualify as a low-luminosity LINER. Galaxies with especially bright nuclei that exhibit large Doppler shifts—and hence rapid motions of the line-emitting gas—are known as Seyfert galaxies after Carl Seyfert, who in the 1940s first identified them as being unusual. Both the LINER and Seyfert types of nuclear activity appear to be the exclusive province of spiral galaxies. By contrast, giant ellipticals are the favored hosts of radio-bright jet activity.

We now know that the quasars are the most powerful of the AGNs—with luminosities as high as 100 trillion Suns, the equivalent of 1,000 Milky Way galaxies. The high redshifts of their optical line emissions first indicated that these point-like sources are at great distances and hence must be extremely powerful. In high-resolution images taken with the Hubble Space Telescope and with the latest groundbased telescopes, we can now see the surrounding—much fainter—galactic structures. Such correspondence leads to the inevitable questions: Has our Milky Way galaxy ever gone through a quasar phase, and could it "go quasar" in the future? We will explore these and other questions relating to the AGN phenomenon in Chapters 6 and 11.

4

THE

MISSING

MASS

Not many years ago one of the more secure fields in astronomy was the study of galaxian masses. Good methods had been developed to measure the masses of galaxies, extensive collections of measurements had been made, and astronomers had lists of values that almost everybody believed. A few worrisome problems arose in the 1960s, particularly regarding the masses derived from looking at the velocities of galaxies in clusters, which seemed to come out too big. But generally it was felt that such "straightforward" problems as the mass of the Milky Way or of the Andromeda galaxy were solved.

By 1980, however, matters had taken a surprising turn, leaving astronomers at present wholly baffled by the problem of galaxy masses. None of the past answers seems to have been correct, because of a completely unexpected and still not understood complication. Before delving into that mystery, however, let us consider the basic methods that astronomers have devised to "weigh" the galaxies.

Weighing Galaxies Star by Star

It is possible to estimate the total mass of a galaxy by adopting very simple assumptions and by measuring things that are easy to measure. For example, the mass of our own Galaxy can be estimated just from its known radius and the number of stars found near the Sun. The simple but not very accurate assumptions that we live in an area of typical star density and that the Milky Way galaxy is roughly a sphere will do the trick.

If we count up the stars in the Solar Neighborhood and add the mass of gas and dust, we find a density of about 3/1,000 solar masses per cubic light-year. The radius of our Galaxy out to the Sun's orbit is about 27,000 light-years, so the volume, if we assume it to be a perfect sphere, is about 90 trillion cubic light-years. The total mass in the sphere is just the volume times the density, so our approximation gives a value of 270 billion Suns. This answer is surprisingly close to the values found by more careful means. The density of stars is actually highly variable in our Galaxy, and they are certainly not arranged uniformly in a sphere. Nevertheless, the simple act of counting up the stars near us one by one and generalizing the local density to that of a much larger spherical volume gives a fortuitously good first approximation and illustrative demonstration of the immensity of our Galaxy's mass.

Weighing Galaxies by Their Orbital Motions

A much better method of finding a galaxy's mass is based on the rotational motion of the galaxy. The method is not much more complicated than calculating the mass of the Sun based on the orbital speeds of the planets. If the Sun were more massive than it is, our Earth would have to go faster around in its orbit or it would be pulled into the Sun. A less massive Sun with less gravity would mean that the Earth should be going more slowly than it does, or else it would fly off into deep space. Thus the speed of the Earth in its orbit matches exactly what it must be for a stable orbit around a star that has one solar mass.

Similarly, the Sun and other stars orbit around the center of our Galaxy at speeds that are determined by the galactic mass. If the speed is measured, and if the size of the orbit is determined, then the mass that regulates the orbit can be calculated. But there is one complication. In the Solar System, virtually all of the mass is in the Sun, which lies at the center of the system. In a galaxy, however, the stars are spread out, and so there is still considerable gravitational attraction on most stars from the mass that lies outside (as opposed to inside) their orbits. Therefore, to determine the total mass of a galaxy, one has to measure the speeds of the outermost stars—where almost all the galaxy is inside the star's orbits.

What an astronomer typically does is to measure the velocities of the stars or other material (commonly the excited gas, because its emission lines and their Doppler shifts are easiest to measure accurately) all the way from the

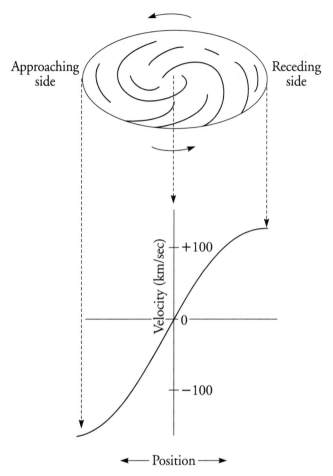

FIGURE 4.1 The rotation curve of a galaxy typically rises from zero in the center to velocities of 100 kilometers per second or more in the outer parts.

center to the edge, looking for the point at which the velocities are responding only to the interior mass. Beyond that point, the rotation velocities steadily decrease according to the square root of the distance, such that the speed drops by a factor of two for every quadrupling of the radius. Such behavior is known as Keplerian motion, named after Johannes Kepler, who found the relation between the planets' speeds and their distances from the Sun in 1609—a discovery that led Sir Isaac Newton to formulate the law of gravity some 50 years later.

The innermost parts of spiral galaxies have rotation velocities that get larger with increasing distance from the center (see Figure 4.1). At greater radii, the velocities begin to level off. They are then expected to finally turn over, becoming smaller again. Once the velocities become Keplerian, the measurements would reliably indicate the total amount of mass contained in the galaxy. Typically, astronomers fit the entire set of velocities, measured at all positions, to various mass models of galaxies, thereby learning something about how the mass is distributed as well as what the total is.

This kind of analysis was exploited quite extensively in the 1960s. Astronomers measured the masses of many galaxies and found a relationship between the luminosity of a galaxy and its mass, and between the Hubble type and its mass. Generally, galaxies of types Sa and Sb were found to have larger masses per unit of luminosity than those of Sc and Irr types—consistent with the higher mass-to-light ratios that characterize the cool and dim stars in early-type systems. In all types of galaxies it looked as if the velocity curve turned down near the limit of the observations. Nature appeared to build galaxies so that we could just barely see, among their outermost stars, the beginnings of Keplerian motion. The curves fit the mass models well, and the distribution of matter in galaxies looked quite reasonable.

Patterns of Motion in Elliptical Galaxies

The method of using the circular orbital velocities of stars does not work very well in a galaxy whose stars do not orbit all in one plane. Elliptical galaxies, for example, have stars that orbit around the center in all planes, frequently with highly eccentric orbits (see Figure 4.2). There is no flat disk, though there is, at least sometimes, a preferred direction of very slow general rotation. The only ready way to measure the mass of such a galaxy is to examine the spread of velocities that the stars show. The speed with which a cloud of stars is buzzing around the center will be sensitive to the total mass, just as the speeds of stars with more circular orbits are. The larger the mass, the larger the average speed of the stars. Some stars will be coming toward us, some going away from us, and some moving across our line of sight. Those motions along the line of sight yield a spread of radial velocities, whose corresponding Doppler shifts will be made manifest in the spectrum of such a galaxy (see Figure 4.2). The width of this spread, called the velocity dispersion, can be used to find the mass of the galaxy, when combined with information on the distribution of stars.

As an example, consider the elliptical galaxy M32, the small, bright companion to the giant spiral M31. It was one of the first to have its mass determined in this way. The Palomar astronomer Rudolf Minkowski measured the velocity dispersion of its stars in 1952, finding a width of 100 kilometers per second. When the velocity dispersion is combined with the size of M32,

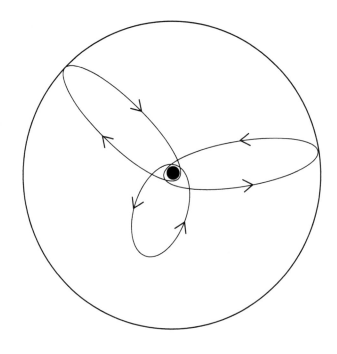

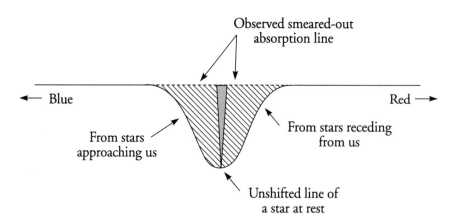

FIGURE 4.2 The patterns of motion in elliptical galaxies. Top: Orbits can be highly elongated. Bottom: The resulting dispersion in line-of-sight velocities smears out the spectral lines of the stars because of the Doppler effect.

one can estimate from this value that the mass must total about 2 billion Suns. In the last 50 years, this method has been refined and applied to hundreds of elliptical galaxies. The observed velocity dispersions range from a few kilometers per second for dwarf ellipticals to about 400 kilometers per second for the most massive giant ellipticals.

Once the mass of an elliptical galaxy has been determined with the velocity dispersion technique, the result can be compared with the galaxy's luminosity, and this comparison in turn can tell us something about the kind of matter the galaxy contains. Astronomers commonly express a galaxy's mass-to-light ratio in terms of the Sun, an average star. Thus a galaxy made entirely of Suns would have a mass-to-light ratio of exactly 1. Spiral galaxies, it was thought, mostly had mass-to-light ratios of just about that value. Elliptical galaxies, on the other hand, seemed to have values in the range of 5 to 10, meaning that these galaxies appeared to contain lots of matter that is fainter than the Sun relative to its mass. Large numbers of dim, lower-mass stars, such as red dwarfs, could easily provide the high mass-to-light ratios measured in these systems. While this may indeed be the case in elliptical galaxies, there are other possibilities, including a rather alarming one that is discussed later on in this chapter.

The velocity dispersion technique can also be used to learn something about the stars in the central bulges of spiral galaxies. One cannot learn anything about the total mass of a spiral in this way, but the mass-to-light ratio can be explored. In M31, for example, the central velocity dispersion is found to be about 160 kilometers per second. The implied mass-to-light ratio is very similar to that for the stars in elliptical galaxies, about 10 times that of the Sun. It seems reasonable to suggest, therefore, that the kinds of stars in M31's bulge are not very different from those that inhabit M32 and other elliptical galaxies. Further evidence tells us that some important differences exist between spiral bulges and ellipticals, resulting from the different chemical and star-formation histories of the galaxies, but these are mostly small effects. In general, it would be difficult to distinguish between elliptical galaxies and the central bulges of spirals if we did not see the outer structure of the spirals.

While considering the dynamical masses of elliptical galaxies, we are inevitably drawn to the strange dynamics of these seemingly bland realms. As previously noted, the stellar orbits in elliptical galaxies are very eccentric and occupy all the different directions, with no preferred plane (this latter characteristic is called isotropy). Until recently astronomers thought that the perfectly round elliptical galaxies were made up of stars with perfect isotropy, no rotation, and a perfectly spherical distribution about the center. Elliptical galaxies that show a more flattened appearance, on the other hand, were thought to be rotating slowly and shaped by their rotation into a flying saucer shape. To the consternation of the model builders, however, the first actual measures of rotation in the 1970s showed that almost no rotation exists in many "squashed" elliptical galaxies. Some show a slow, gradual rotational motion, but nowhere near enough to explain the galaxies' flattened shape.

One possible interpretation of this surprising result is that the galaxies are prolate instead of oblate. A prolate galaxy is shaped like a fat sausage instead of like a thick hamburger; its shape has nothing to do with rotation in a preferred plane. Another possibility is that the galaxies are oblate all right, but the shape is not simply a result of rotation; it is rather some mix of motions that existed when the galaxy formed and that now show up in the distribution of orbits (which for the "flattened" ellipticals are preferentially concentrated to a particular plane). The still unexplained flattening of elliptical galaxies reminds us that objects as huge as galaxies can be much more complex than they look, even when they seem to be pure, simple, and featureless (see Chapters 2 and 9).

Motions of Double Galaxies

For those galaxies in double systems, another method of determining the masses can be applied. Two galaxies that revolve around each other also have to obey Newton's law of gravity. Once again, the orbits and speeds must depend on the gravitating masses. By looking at just one double galaxy, we have no hope of deriving the masses—the orbital periods are millions or billions of years, much too long for us to wait. Furthermore, we can only see the pair from this one direction, so we cannot determine the true angle that each galactic orbit makes with our line of sight. These problems can be overcome, however, if we look at lots of different double galaxies and solve for their properties statistically. Even though we cannot follow the orbits of any given pair, we can look at enough pairs to build up an averaged estimate of their masses.

The average mass that is deduced will be somewhere in between the actual masses, and hence not representative of either. Nevertheless, one can accurately compute the mass-to-light ratio for the system as a whole. It is then a simple matter of multiplying this ratio by an individual galaxy's luminosity to obtain a good estimate of that galaxy's individual mass. In actual practice this must be done for many pairs of elliptical galaxies to compensate for the various unknown angles and orbital shapes.

The results of studying pairs of all different types of galaxies are rather surprising. Instead of finding mass-to-light ratios of 1 to 10—the range for individual galaxies analyzed by the methods previously described—astronomers have obtained much larger numbers. Typically, elliptical pairs give values of about 75 and spiral pairs range in the neighborhood of 20 to 40. These values puzzled the people who found them, and were so different from what had been expected that a concerted effort was made to find out how they might have gone awry.

Were the assumptions wrong in some way? Perhaps galaxies in pairs are inherently heavier (for their brightness) than lone galaxies for some evolutionary reason. Or perhaps the statistical approach has a flaw in it somewhere. Because of these doubts, astronomers tended to be cautious about the answers that came from double galaxies. They should not have been, but should have saved their worries for the more traditional methods. As the next sections show, the evidence now suggests that the double galaxies were giving better answers than astronomers realized.

Motions of Groups and Clusters

Galaxies in general tend to exist in groups; they congregate. Some, like the Milky Way, belong to small organizations like the Local Group, while others are members of huge clusters, containing thousands of galaxies (see Plates 28 and 29). In all cases, this fact provides us with another method of finding galactic masses. In a cluster of galaxies, all are moving in accordance with the gravitational pull of the others. How fast they move on the average depends on how far apart they are on the average and how massive they are. The situation is similar to that of the velocity dispersion of stars in a galaxy, but now we are considering the motions of individual galaxies in a cluster.

If it is assumed that the clusters of galaxies are stable—that they are not falling into themselves or flying apart—then the motions and separations of their members should give us a measure of their masses. Such an assumption is reasonable, because the dissolution time of galaxy clusters in the absence of gravity is on the order of a billion or so years. Without gravitational stability, we would not be witness to any of the clusters that we see today.

The problem with the cluster method was that it, too, seemed to give the wrong answer. When average mass-to-light ratios were first calculated in this way in the early 1960s, the results were astonishing. Instead of values around 1 to 10, the answers came out in the 100s and even in the 1,000s. How could this method be so wrong? Various suggestions put forward included the possibility that the clusters are expanding, that they are contracting, that they consist of anomalously massive galaxies, that there are lots of double galaxies in clusters (which would lead to higher measured velocities), or that there must be lots of intergalactic matter between the galaxies in clusters—enough to swamp the gravitational field of the galaxies themselves.

We now feel a little more cheerful about the cluster results than we did at first. No doubt, all of the aforementioned possibilities enter in to some extent, but the major explanation is something quite different. All along, the galaxies have been hiding a disturbing secret from us: that they are full of mysterious dark matter.

The Discovery of Dark Matter

Knowledge comes to us in various ways, of which the most exciting are the rare breakthroughs. These momentous events occur after scientists have been stuck for some time and realize that something is missing; some vital bit of knowledge is there on the threshold, but remains elusive and unfound. The study of galaxy masses went through a phase of this sort when most astronomers felt that there was something wrong with the field, that some important fact had escaped them. The results that were coming in from various ways of measuring masses did not agree, and the problem was especially acute for clusters of galaxies. The field definitely needed a breakthrough.

The Swiss astronomer Fritz Zwicky was the first to measure and draw attention to the high radial velocity dispersions in galaxy clusters. Working at the California Institute of Technology, he concluded as early as the 1930s that galaxy clusters have anomalously high mass-to-light ratios. In describing this puzzling result, he was the first to coin the term "missing mass." As cantankerous as he was brilliant, Zwicky was unable to convince most astronomers of his radical findings. It turns out, though, that he was the first harbinger of the coming revolution.

The first independent indication that a breakthrough was imminent was a study done in the 1970s by Vera Rubin and Kent Ford of the Carnegie Institution. Using a new electronic spectrograph, they obtained spectra along the major axes of several nearby spiral galaxies, including our neighbor M31. The resulting Doppler-shifted spectral features yielded velocity curves that stayed persistently high, with some modest downturns (and some upturns) at the farthest observed points. Rubin and Ford concluded from their "flat" rotation curves that something massive and dark was out there, keeping the orbital velocities elevated.

Complementing this optical study, radio astronomers had begun to obtain well-resolved observations of nearby spirals at the 21-centimeter wavelength emitted by neutral hydrogen. In most of the spirals, they were able to detect hydrogen gas well beyond the optical limits. Even at these great distances from the galactic nuclei, the rotation curves refused to turn down and become Keplerian (see Figure 4.3). Where the optical data had suggested that a downturn had been reached, the newer data from the neutral hydrogen showed the velocities remaining fairly constant.

Such high orbital velocities so far from the galactic centers could persist only if there existed large amounts of mass in some invisible halo, far beyond the visible limits of the galaxies. All kinds of things were thought of to account for this unseen mass. Perhaps it is in the form of very dim red stars, it was suggested, or maybe gas that is ionized so that we cannot see it as neutral hydrogen. But these simple suggestions, like others involving familiar objects, were soon ruled out by various careful observations. The mass out there could not be anything that simple.

Meanwhile, other evidence began to turn up indicating that massive halos of invisible matter might be common among galaxies. More sophisticated theoretical models seemed to demand that there be some very massive halo in order to keep the observed flat disk of a spiral galaxy stable. It was argued that the disk would disintegrate unless held in place by the overwhelming gravity of an enveloping mass.

By the 1980s it began to look as if there were no galaxies whose mass was primarily located within the visible disk. Now we find that a few galaxies do show a Keplerian curve in their outer parts, but the majority do not. Most of the optical and radio rotation curves seem to go on at about a steady velocity right out to the outermost detectable point, even when the most powerful modern techniques are used to record the faintest possible light. If basic Newtonian gravity is correct, we must conclude that most galaxies are dominated by matter that extends well beyond their observable limits. Indeed, the problem is not that the mass is "missing," it is that the mass is there and yet remains completely invisible!

If galaxies truly have massive dark halos, then the strange disagreements discussed above can be understood. The rotation curve method gives us only the amount of mass found inside the outermost measured point (roughly speaking), and the velocity dispersion method tells us only about the mass-to-light ratio in the center, which makes it necessary to extrapolate outward using the light distribution to find the total mass. Neither method would detect a more extensive invisible halo beyond the visible confines of the galaxy. But the double galaxy method would, because the galaxies revolve around each other in orbits that are largely or entirely outside the massive halos of

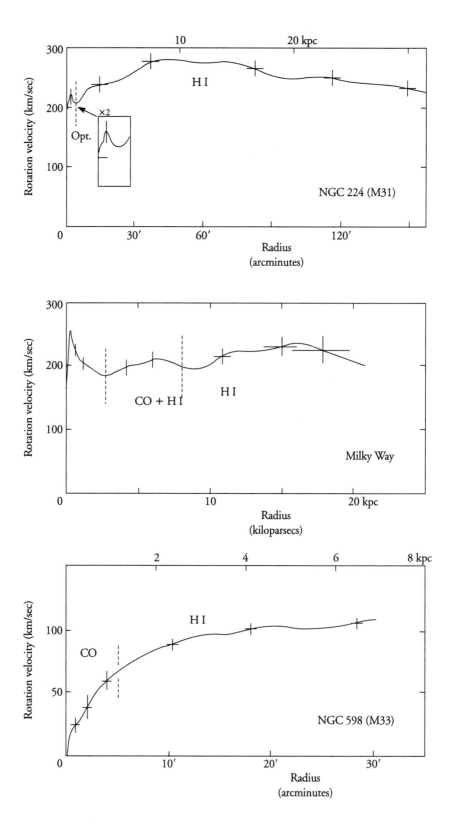

FIGURE 4.3 (facing page)   The rotation curves of spiral galaxies do not decline at large distances from the centers—contrary to expectations based on the radial falloffs of star-light. These nearly flat rotation curves indicate large amounts of unseen (dark) matter gravitationally binding the rapidly rotating CO and H I gas. Here, the galactocentric radius is expressed both in arcminutes of angle, where one arcminute is 1/60 of a degree, and in kiloparsecs, where one kiloparsec (kpc) is 1,000 parsecs, each parsec (pc) being 3.26 light-years.

the individual members. Similarly, the cluster method should accurately gauge the total mass of the clustered galaxies along with any material that lies between the galaxies.

Within the clusters, dark matter may constitute more than 90 percent of the total gravitating mass. This matter should have a strong effect on any light rays that happen to traverse the cluster. Recent deep imaging of rich galaxy clusters has shown that they indeed operate like gigantic gravitational lenses, distorting the observed shapes of more distant galaxies and quasars into bizarre arcs and rings (see Plate 29).

One of the most compelling evaluations of gravitational lensing by dark matter in a cluster was performed by J. Anthony Tyson and his colleagues at Lucent Laboratories (formerly Bell Laboratories) in New Jersey. Working with a deep Hubble Space Telescope (HST) image of the Abell cluster CL0024+1654 that showed multiple gravitationally lensed images of a more distant galaxy, Tyson's team was able both to reconstruct the shape of the lensed galaxy and to derive the spatial distribution of gravitating matter in the foreground cluster (compare Figure 4.4 with Plate 29). The resulting distribution shows a grand pile-up of dark matter that peaks at the center of the cluster. The visible galaxies add discrete peaks to the overall pile-up but are overwhelmed by the ubiquitous dark matter.

Similar distributions are evident in X-ray images of rich galaxy clusters (see Figure 4.5). In such clusters the hot intra-cluster gas traces the overall gravitational field of the cluster. Moreover, the size and temperature of the X-ray emitting gas can be used to derive the total mass necessary to bind the gas to the cluster. Once again, the mass determinations are staggering—yielding mass-to-light ratios of several hundred. Even after we account for the considerable amounts of hot gas in these X-ray bright clusters, 90 percent of the remaining mass is still unexplained.

The distressing aspect of these new measurements is that if the large-scale dominance of dark matter is true, then astronomy as it is currently practiced does not detect the majority of the Universe. Most of the matter in the cos-

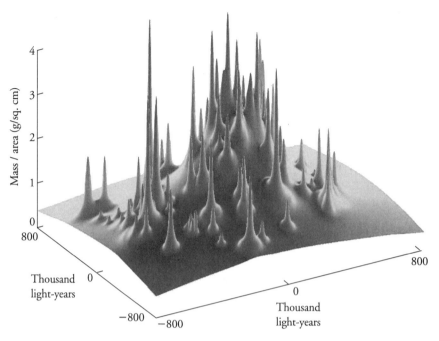

FIGURE 4.4 The distribution of gravitating matter in the rich Abell galaxy cluster CL0024+1654 as derived from the cluster's observed gravitational lensing of a background galaxy (see Plate 29). The peaks indicate individual galaxies, while the smoothly distributed matter is for the most part dark.

mos is locked up in some unknown form in the halos of individual galaxies and in the spaces between clustered galaxies. What we see as galaxies are merely the tips of some very large icebergs.

Perhaps we are fixating too much on the troublesome dark matter and should be thinking of alternatives. One controversial alternative is to alter Newtonian gravity on large scales. Proponents of the so-called MOND theory (short for Modified Newtonian Dynamics) contend that the relation between gravitational force and acceleration is different on large scales—where the acceleration is typically weak. The upshot is a greater degree of motion for a given amount of gravitating mass, which explains the anomalously high velocities that are observed at large radii within galaxies and on larger scales within galaxy clusters. This dynamical modification, however, remains unmotivated by any underlying physical theory. Moreover, it fails to explain the gravitational lensing effects produced by galaxy clusters that otherwise indicate large amounts of gravitating dark matter. For now, we had

 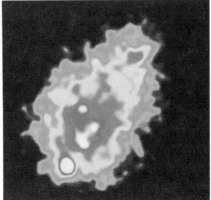

FIGURE 4.5   Hot gas in the galaxy cluster Abell 1367. On the right is the distribution of X-ray emitting "coronal" gas. Comparison with the visible-light image of the cluster, on the left, reveals the ubiquitous nature of the hot gas. Its size and temperature provides a dynamic measure of the cluster's total mass—amounting to more than 10 times the mass evident in luminous stars and gas.

best face up to the probable existence of dark matter, and to the challenges that it presents.

What Is the Matter?

Many kinds of things have been proposed to explain the unseen matter in galaxies and galaxy clusters. When physicists first suggested that the tiny particle called the neutrino might actually have a small mass (it had previously been believed to be massless at rest), someone immediately said that the halos might be made of neutrinos. When it was reported that physicists had detected a monopole (a single isolated magnetic pole) with a tiny mass, someone else immediately suggested that the halos might be made of monopoles. As each new possibility arose, there always seemed hope that it might explain galactic halos.

We now recognize that even if neutrinos have mass, they are far too zippy to be constrained to the halos of galaxies. These and other forms of hot dark matter (HDM) are better suited for much larger structures spanning millions of light-years. As for the monopole, the one detection to date may have been a fluke. We are left with a rather thin list of improbable objects, each of which comprises some form of cold dark matter (CDM), none of which seems to make a lot of sense, alas. On the list is everything people have been able to concoct that has mass, can be bound to a galaxy's halo, and is invisible.

MACHOS and WIMPS

The most familiar CDM candidates are known as MACHOS—short for massive compact halo objects. These are made of normal subatomic stuff, such as protons and neutrons, but are somehow immune from emitting or absorbing detectable amounts of electromagnetic radiation. Planets like the Earth, if unaccompanied by a shining star, would have mass but would emit too little light to be detected. Smaller objects, boulders or pebbles, for instance, would also qualify. The problem with such things is that no one can think of a way to produce them in the necessary vast numbers. We are rather sure that a planet cannot form without a star nearby and the same goes for boulders.

Larger objects—up to 1/100 the mass of the Sun—would be similarly faint. Such objects are known as brown dwarfs, and have been the subject of several intense surveys recently. The numbers of brown dwarfs that have been found via these various search techniques are significant but are far from explaining the bulk of the dark matter. The same high numbers but insufficient net masses also seem to hold for white dwarfs, the end products of intermediate-mass stellar evolution.

After white dwarfs, the only remaining dark objects of stellar pedigree are neutron stars and black holes, which have mass and virtually no light, and which might have been preferentially formed in the outer parts of proto-galaxies. According to this picture, each galaxy we see today would have gone through a spectacular epoch of starburst activity, spawning vast numbers of high-mass stars that in turn collapsed into neutron stars and black holes.

Although an early starburst epoch may have been common to many large galaxies (see Chapter 16), the intensity required to form sufficient numbers of these stellar remnants is difficult to reconcile with the degree of chemical enrichment that we see today. From surveys of massive stars and supernova remnants in our own Galaxy, for example, we find that most massive stars must die in supernova explosions. Therefore, the creation of each neutron star and black hole remnant probably involved the explosive injection of heavy elements into the interstellar medium. This sort of chemical enrichment, while evident in our Galaxy, is not nearly at the levels that would arise if our dark halo was laden with black holes and neutron stars.

For now, it seems that MACHOS are unable to account for the dark matter. Other candidates fall under the more exotic rubric WIMPS—short for weakly interacting massive particles. Neutrinos are certainly weak interactors; several hundred billion billion (10^{20}) neutrinos per second are currently streaming through each of our bodies, with nary an impact. But they have extremely low mass—if any—and hence have unbridled speeds close to that of light. Candidates that are weakly interacting and are sufficiently massive and slow to settle into galaxy-size clumps include some of the most bizarre particles in the subatomic zoo.

According to theorists pursuing the holy grail of grand unification between all particles and forces, clumpable WIMPS include photinos (massive counterparts to the massless photons), gravitinos (massive counterparts to the massless gravitons), and other partinos that have been invoked to provide supersymmetry among particles of integer spin (bosons) and half-integer spin (fermions). For example, the photon is a boson of spin 1, so according to the theory of supersymmetry, the photino is predicted to be a massive fermion of half-integer spin. Unfortunately, despite intensive searches with the most powerful particle accelerators, none of these objects has yet to be found. WIMPS remain the exclusive province of theory.

As the new millennium unfolds, we are left speculating about the nature of the dark matter. Whatever it is—rocks, black holes, or exotic subatomic particles—the disturbing possibility remains that most of our Universe is hidden from us. We live in a deep, darkly ponderous cosmos, lit only by an occasional candle.

5

CREATION AND

EVOLUTION

One of the greatest goals of modern astronomy is to understand how galaxies were formed and how they evolve. When Edwin Hubble was classifying galaxies in the 1920s, it was tempting to believe that the classes of galaxies represent different stages in their development. But that idea turned out to be mostly wrong. Nowadays, the challenge of reconstructing the life histories of galaxies has led astronomers to ponder how galaxies came into being in the first place.

The nature of the Universe before galaxies existed is not well known, and the hypothetical characteristics one attributes to it depend to a great degree on the cosmological model one chooses (see Chapters 14 and 15). Most currently accepted cosmological models posit a general expansion from some singular moment, generally known as the Big Bang, when the density and temperature of the primeval Universe were both extremely high. The physics that describes the expanding space-time in these models can be followed fairly reliably right up to the epoch when density and temperature were low enough to permit galaxy formation.

It took about a million years for things to decompress and cool off enough for matter to become important in the Universe. Before that time, radiation (intense light) predominated and chunks of matter such as stars or galaxies could not have formed. But when the temperature was about 3,000 K and the density was about 100 atomic nuclei per cubic centimeter (conditions remarkably similar to those in H II regions), neutral atoms of matter could form at last. Once the positively charged nuclei and negatively charged electrons were bound into neutral atoms, the matter could decouple from the radiation and begin to clump up gravitationally. At that time, the nuclei of hydrogen and helium were just about the only things available to form into atoms and collect into protogalaxies. Later generations of stars would be required to cook up most of the heavier elements that characterize our life on Earth.

Although one can imagine several mechanisms for forming galaxies out of this hydrogen and helium gas, finding even one model that will actually work under the probable conditions in the early Universe is difficult. For an isotropically expanding, smoothly distributed universe of uniform temperature, there is very little reason for a galaxy to form; such an idealized universe will never have a galaxy in it. The undeniable existence and prominence of galaxies in our Universe, however, argues that the pregalactic environment was nothing like a perfectly uniform gaseous cloud. Instead, there had to have been irregularities, of one sort or another. But what kinds of irregularities, and where did they originate?

Fluctuation

Most theories of galaxy formation begin with some sort of "seed" in the cosmic firmament, from which galaxy-size realms could grow. These seeds usually take the form of small fluctuations in the prevailing density at the time. The origin and detailed nature of the fluctuations remain uncertain. Currently favored candidates include possible quantum fluctuations during a putative inflationary epoch, imbalances in the ratio of matter to antimatter, cosmic strings, and other "defects" in the topology of space-time (see Chapter 15). Although we know very little about the primordial fluctuations, we do know that they were very small. Observations of the cosmic microwave background radiation—which traces the epoch when matter began to dominate over radiation—shows irregularities no greater than a few parts per 100,000 (see Plate 32). A freshly polished ice rink comes close to replicating the cosmic background's texture. How such minuscule density fluctuations amplified into the galaxian constructs that we see today continues to challenge astronomers and cosmologists.

Perhaps we are looking on the wrong size scales. The Cosmic Background Explorer (COBE) satellite—launched in 1989—was able to map the fluctuations on scales of several degrees, but no finer. The corresponding size scales at the distance of the cosmic microwave background would be roughly a billion light-years—bigger than the largest galaxy clusters we see today. More recent measurements on finer scales, using microwave telescopes atop the snow-fields of Antarctica and aboard high-flying balloons, suggest higher amplitude fluctuations on equivalent scales of about 100 million light-years. Theoretical investigations of the very earliest Universe predict the spectrum of fluctuations to peak at similar levels and size scales. These sorts of seeds make life a little bit easier for the galaxy builders, but not by much. Many cosmologists have pinned their hopes on NASA's Microwave Anisotropy Probe (MAP) and the European Space Agency's PLANCK satellite (named after the pioneering quantum physicist Max Planck), whose higher-resolution mapping of the cosmic microwave background will provide critical information on the smaller-scale fluctuations.

Perhaps we are worrying about the wrong sort of matter. If the Universe is bathed in cold dark matter (CDM), and especially if weakly interacting massive particles (WIMPs) dominate the CDM, then the broad but low-amplitude fluctuations we see might be about right. According to this scenario, the WIMPs would have coagulated first and then attracted the baryons, heavy elementary particles including protons and neutrons. At the epoch of the cosmic background radiation, we would then be seeing this latter stage, where the photon-scattering plasma of charged baryons and associated electrons is just beginning to concentrate into the WIMPy cores. Following this epoch, we are confronted with a ghostly "dark age," with nothing in the form of luminous stars or recognizable galaxies to guide our way.

Instability

Most early attempts to find ways by which the material in the Universe could have condensed into galaxies were based on an idea first introduced in 1902 by Sir James Jeans. Although we now assume that the early universal gas expanded according to a relativistic cosmological model, Jeans's ideas were based on a simpler, Newtonian universe. Using classical Newtonian mechanics, he showed that a gravitational instability occurs when a pocket of denser material (called a perturbation) becomes sufficiently dense and large.

The characteristic size of the density perturbation that is just barely unstable is called the Jeans length. It increases in direct proportion to the characteristic speed of the particles in the medium (often known as the sound-speed) and decreases according to the square root of the density. At the epoch of "decoupling" between matter and radiation, the Universe had a density of a few hundred atoms per cubic centimeter, a temperature of 3,000 K, and a corresponding sound-speed of 5 kilometers per second. The resulting Jeans length was roughly 50 light-years across.

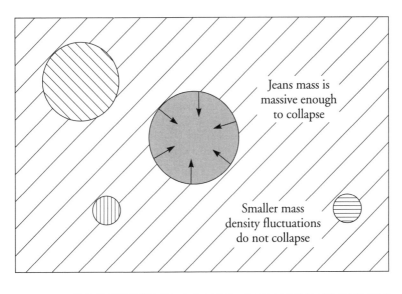

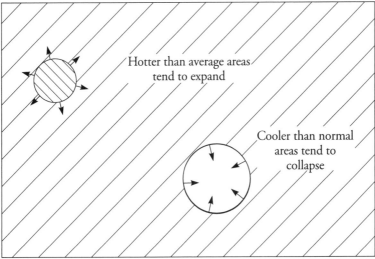

FIGURE 5.1 Primordial instabilities. Top: Gravitational (Jeans) instability. Above a limiting mass, known as the Jeans mass, overdense regions are gravitationally unstable and collapse under their own weight. Bottom: Thermal instability. Cooler areas have lower pressures than their surroundings, and so tend to collapse. The denser material is able to radiate and cool at a greater rate, thereby reducing the temperature and pressure further, which induces further collapse, and so on and so forth.

The Jeans mass is defined as the minimum mass of material that is gravitationally unstable and hence amenable to contracting under its own weight (see Figure 5.1). It can be easily computed by converting the Jeans length into a volume (by taking its cube) and multiplying that volume by the density. At the beginning of the "matter era," the Jeans mass was about 50,000 solar masses; so at that point in the history of the Universe, perturbations with masses larger than this value would have been unstable and would have contracted. The Jeans mass limit at least partly explains the lack of galaxies with masses less than 100,000 Suns.

It is not possible with the simple Jeans model to examine the situation during the "era of radiation" because the effects of radiation pressure on the ionized gas during those times are not included in this simple analysis. But several astronomers and cosmologists have examined the more complicated situation with radiation present, and the results roughly agree with the simpler models. These investigations have also provided some physical basis for the upper limits that we find for the masses of galaxies (10^{13} solar masses) and of galaxy clusters (10^{17} solar masses).

Another important mechanism for condensing matter is thermal instability. Regions of slightly higher density radiatively cool off more rapidly than their surroundings. The hotter regions around them compress them even further, increasing their density. Thus a small density perturbation can become increasingly more unstable (see Figure 5.1). In some instances, the amount of cooling depends on the square of the density, an effect that amplifies the thermal instability's compressive effect.

Thermal instabilities may have worked in tandem with Jeans's gravitational instability to produce the first protogalaxies following the epoch of decoupling. Before that epoch, the radiation would have dampened out any irregularities produced by these instabilities. It is possible that extremely large-scale irregularities in the ponderous dark matter may have survived the damping, however, and if so that would explain the galaxy clusters and superclusters we see today.

Contraction

After individual protogalaxies have achieved their gravitational identity through some form of instability in the pregalactic gas, they collapse to form galaxies of much smaller dimensions and higher densities, leaving the intervening space nearly empty. The actual means by which the contraction takes place is something that we can explore only by theoretical modeling. We have not yet discovered a nearby galaxy that we can say for sure is very young compared with the estimated age of the Universe (although some promising protogalactic candidates have been identified). Therefore, we have had to imagine the precontraction environment from the observed characteristics of nearby galaxies and from what we can learn of the past by peering deeply into the distance (see Chapter 16).

A nonrotating protogalactic cloud	A rapidly rotating cloud	A slowly rotating cloud
collapses	collapses	collapses
becoming a black hole	to a thin rotating disk	to a slowly rotating spheroid

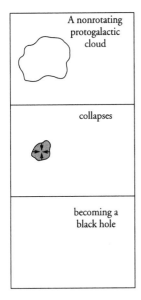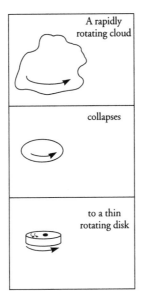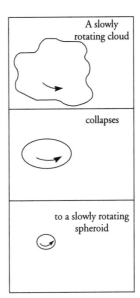

FIGURE 5.2   Collapse scenarios. Left: The fate of a nonrotating protogalaxy, where the gas does not condense into stars, but instead collapses into a super-massive black hole. Middle: The fate of a rapidly rotating protogalaxy, where the collapsing gas collisionally dissipates into a thin rotating disk. Right: The fate of a slowly rotating protogalaxy, where the gas rapidly condenses into stars with a spheroidal distribution of orbits, that is, an elliptical galaxy.

We can also approach the problem by proposing plausible initial conditions and carrying through the calculations to see whether we can arrive at a realistic picture of a contracted galaxy. The initial conditions that we must consider for these calculations include the mass of the galaxy, its angular momentum, its dimensions, its temperature, its chemical properties, its magnetic field, and the turbulent motions within it.`

Consider the simplest initial situation, where the protogalaxy is cold, uniform in density throughout, perfectly spherical, and without turbulence, rotation, magnetic field, or external influence. For an object with a mass comparable to that of the Milky Way—on the order of 10^{11} solar masses—such a set of initial conditions leads to a completely unbraked collapse. The gravitational potential of such an object is sufficiently large that no physical process can stop its collapse into a massive black hole. Calculations indicate that in a cosmically short time, such an object will disappear from our space-time (see Figure 5.2). It passes through the Schwarzschild radius, which according to Einstein's general theory of relativity corresponds to the "horizon" that is established when a massive body contracts to such a small size and enormous density that light can no longer escape from it. Once inside this horizon, the object disappears to an outside viewer, and only its gravitational field is detectable. Thus the very simplest initial conditions would not lead to a galaxy at all.

A more reasonable set of initial conditions would be the following: By one of the processes discussed above, the gas cloud has already contracted to the point where the density is high enough for it to be stable in spite of the expansion of the surrounding Universe, let us say approximately one atomic nucleus per liter of volume in a universe whose mean density is several times lower. If we assume a mass of 10^{11} solar masses, then this density for a spherical cloud implies an initial radius of about 600,000 light-years (as opposed to 100,000 light-years, which would be typical for a galaxy of this mass *after* contraction).

For contraction to occur, the kinetic, magnetic, and gravitational energies must all be appropriately balanced. A typical set of initial conditions for contraction to begin would require that the rotational velocity must be small, less than about 40 kilometers per second, any turbulent velocity must be less than about 30 kilometers per second, the temperature must be less than 2,000 Kelvins, and the strength of the magnetic field must be reasonably small, less than about 0.2 micro-Gauss (0.2×10^{-6} Gauss), where the magnetic field of the Earth is roughly 1 Gauss).

If the density of the cloud remains uniformly distributed during contraction, then the release of gravitational energy rises in inverse proportion to the shrinking radius. The temperature, however, remains approximately the same, as the release of gravitational energy is converted into radiation that then escapes from the system. When the density of the material is so large that it becomes optically thick (opaque) to the photons of light, the cloud is no longer able to radiate away the energy produced during gravitational contraction. Once that happens, the thermal energy of the gas cloud (the amount of energy represented by the motions of the gas particles) rises dramatically with any further decrease in radius. At or near the so-called thermal radius, this thermal energy of the cloud can stop the contraction.

The turbulent energy, which corresponds to the random motions of macroscopic parcels of gas in the cloud, is not as important at these early stages, because it is rapidly dissipated via collisions among the gas parcels. As the matter clumps up in the protogalaxy, however, turbulence becomes an important regulator of further contraction. Indeed, the chaotic distribution of stellar orbits in elliptical galaxies can be approximated in terms of a turbulent energy component.

The magnetic energy, which increases as the cloud shrinks, never becomes larger than the gravitational energy if it started out smaller to begin with.

Eventually, the radius becomes small enough that the rotational energy balances the gravitational energy; this defines the rotational radius. At yet another critical size, stars condense out of the gas, and the gas cloud rapidly transforms into a galaxy of stars. This so-called condensation radius depends on several factors, including the rate at which turbulent energy is dissipated—and star formation catalyzed—by collisions among the gas parcels. The ultimate fate of the contracting cloud depends on the relative sizes of the three critical radii corresponding to the rotational, condensation, and thermal limits. Three interesting possibilities emerge, depending on which is largest.

When the rotational limit is the largest, contraction is stopped by the rotation (see Figure 5.2). The centrifugal forces, however, are limited to the plane of rotation in such a way that contraction continues perpendicular to this plane until a thin disk is formed. The thin disk is conspicuous by its shape and by its rotation; it is well on its way to becoming a spiral galaxy.

In the case where the condensation radius is the largest, star formation begins before the effects of the rotation become important braking factors in the contraction. As the density increases, the rate of star formation becomes greater and most or all of the gas goes into star formation. When the contraction is finally halted at the limit, there is little or no gas left to dissipate energy efficiently. Therefore, a disk does not form. Energy requirements say that the object must then expand somewhat until the radius reaches another critical value. The stars will orbit in such a way that the galaxy will be nearly spherical, depending on the actual amount and distribution of the initial angular momentum. With these properties—the nearly spherical shape, the lack of gas, and the large number of stars formed near the beginning of its history—the object would be classified as an elliptical galaxy (see Figure 5.2).

In the third case, when neither the rotational limit nor the condensation limit is large enough to brake the contraction, the cloud gets smaller and smaller, until finally a super-massive starlike object is formed at the thermal radius. This will probably evolve quickly into a black hole, invisible and almost undetectable (see Figure 5.2).

Accretion

Throughout the foregoing discussion, we have assumed that the protogalaxy contracts as a monolithic unit, having emerged whole from the expanding cosmos. This view is often regarded as the result of a "top-down" approach to structuring the Universe, whereby the largest structures form first—with the galaxies subsequently condensing within them. Such an approach has its problems, because it predicts galaxies to have formed much later than what we actually find. But monolithic galaxy formation is not precluded by our theoretical problems with it. Indeed, deep images of distant galaxies provide support for large galaxies having existed in the early Universe. Somehow, big galaxies were able to get themselves together from the get-go.

One way to ensure quick formation of galaxies is to invoke a "bottom-up" approach, whereby subunits with masses close to the Jeans limit aggregate to build up the galaxies that we are familiar with today. This approach is favored by several theories that consider the clumping of cold dark matter shortly after the decoupling epoch. It is also supported by observations of galaxies (such as the Milky Way) accreting smaller galaxies during the current epoch.

By its very nature, the bottom-up approach handily explains the sprawling galaxies that are seen in deep images, such as the Hubble Deep Field (see Plate 30). But it is less able to explain the large centrally concentrated galaxies that are also seen at high redshifts (and hence in early epochs), quasar activity, and other tracers of super-massive concentrations that are evident during the very early epochs, as well as the massive bulges of ancient stars that we find today in many nearby spiral galaxies. We are left wondering how such incredible concentrations of matter could have formed so quickly out of a hodge-podge of aggregating subgalaxies.

Such primeval monoliths constrain the bottom-up scenario of hierarchical galaxy formation by pushing back the epoch of piecemeal galaxy building to the first billion or so years following the Big Bang. Perhaps we will ultimately find that both scenarios share in the formation of galaxies, in an exquisite balance that has evolved over time. For example, it is altogether plausible that the cores of giant galaxies first formed from a maelstrom of subgalactic clumps, perhaps under the attractive influence of preexisiting cold dark matter "halos." Inner disks could have settled down next, reinforced by their rotation (determined mostly by the initial rotation of the collapsing protogalactic cloud).

As neighboring proto-dwarfs succumbed to the overwhelming gravity of the dominant protogalaxy, successive accretion events ensued, thus building up the outer disks. Today, the outer disks of giant spirals are just beginning to settle down, while the same holds true for the gas-rich dwarf irregulars.

Watching Galaxies Evolve

Once a galaxy has taken shape, it spends the rest of its life in ceaseless transformation—converting its gas into stars and responding to both inborn and externally driven dynamics. Stars form, they die, and they eject material laden with heavy elements, which, in turn, finds its way into new stars. As star formation proceeds, the chemical composition of each new stellar generation becomes increasingly enriched with heavy elements. Eventually, the formation of new stars yields up coteries of rocky planets—at least one of which has played host to the emergence of life and, more recently, intelligent beings capable of investigating the galaxian Universe that spawned them.

We cannot see a galaxy change. A human life is at least 1 million times too short. But we can see the effects of evolution by looking back to earlier and earlier stages in the evolution of our Universe, when the galaxies all appear younger. Light from a galaxy that is 10 billion light-years away, for instance, has taken 10 billion years to reach us, so the image that we see and measure is that of a 10-billion-year-younger galaxy. The concept of lookback time relies on this observed effect of light's finite speed. If the Universe is 15 billion years old (we are not yet sure of the exact figure), then the galaxy as observed at a lookback time of 10 billion years would be only about one-third the age of the galaxies near us, where the lookback time is negligible. Of course, this argument depends upon the belief, which is supported by local galaxies and predicted from cosmological models, that most galaxies contracted and formed at about the same time, shortly after the Big Bang.

To see galactic evolution, we "merely" need to look farther into the distance. The first couple of billion light-years is too small a distance to detect change, but the galaxies beyond that show real differences. Among the many recent discoveries of "unusual" galaxies at early epochs, astronomers have found a strong evolutionary effect in the colors of galaxies at lookback times of about 10 billion years. Using sensitive electronic imagers and the biggest telescopes, they have measured the faint light from distant galaxies with sufficient accuracy to compare with what theoretical models show. Much as the theoretical models predict, galaxies were both brighter and bluer back then.

The calculations of the Yale astronomer Beatrice Tinsley, who devoted much of her brief but creative life to the study of galactic evolution, have helped astronomers to understand the details of these aging effects. From models produced by Tinsley and her coworkers in the 1970s, we know that the rate of dimming and the color changes among galaxies depend on many things: the distribution of stellar masses in each episode of star formation, the history of star formation over time, the chemical enrichment process, and several other considerations. Recent observations of distant galaxies are beginning to give us these details (see Chapter 16).

Another conspicuous difference between the young galaxies in the distant Universe and galaxies nearer the present is that there were many more interacting galaxies with active and erupting nuclei back then. Both quasars and radio galaxies increase in density as we look farther and farther away. Therefore, these tempestuous objects must have been more common in the early years of the Universe. The current consensus of theoretical models maintains that quasar and other nuclear activity is generated by the collapse of supermassive objects, perhaps black holes, at the centers of galaxies. Black holes are harmless enough if there is nothing to drop into them, but become agents of violent energy if gas or stars come too close to their gravitational field. Interacting young galaxies, still full of unprocessed gas, could have fed this material into their central cores at a much greater rate than is observed in most galaxies today. If black holes were lurking there, these galaxies would have flared up more frequently as quasars or radio galaxies (see Chapter 11). Now, it seems, most of these violent fireworks have died down.

Despite our ephemeral lives as humans, we can learn about galaxian processes that took billions of years to occur. We do it by looking billions of light-years into the distance, and so witnessing events that transpired billions of years ago. In Chapters 15 and 16, we will take a more detailed look at the formative and primeval epochs of galaxian history—and attempt to bridge the gap between those remote epochs and the present day. In the next section, we will carefully examine our current epoch by exploring the resplendent realms closest to us—beginning with our own Milky Way galaxy.

II

NEARBY GALAXIES

6

THE

MILKY

WAY

Consider yourself fortunate if you have experienced the Milky Way with your own eyes. In polls of our undergraduate students, we find that less than a quarter have ever seen the ghostly glow of the Milky Way. Artificial lighting is to blame—a seemingly irreversible juggernaut of municipal, residential, and commercial illumination that has all but vanquished the diffuse splendors of the night sky. It is enough to make a grown astronomer weep.

To see the Milky Way nowadays, we must leave the cities and suburbs behind and head for darker skies. From mountaintop aeries, whispering deserts, secluded seashores, and open seas, we can still bear witness to those clear moonless nights when the Milky Way looms large and majestic across the firmament.

First Impressions

As recently as 200 years ago, such sights could not be denied. Every preindustrial culture had to contend with the reality of the Milky Way as much as it did the wandering planets. For those living in the floodplains of ancient Egypt, the Milky Way and other marvels of the sky were presided over by the sky goddess Nut. She is shown in a bas-relief sculpture sprawling over the disk of the world (ruled by the god Geb), her feet planted on one side of the world, and her extended hands touching the opposite horizon (see Figure 6.1). The Milky Way itself was thought to be a heavenly analogue of the Nile

FIGURE 6.1   This carved sarcophogus lid depicts the Egyptian sky goddess Nut vaulted over the disk of the world, itself ruled by her mate, the god Geb. Myriad stars are evident along Nut's torso and in an arc below her. The carving dates from the fifth century B.C.E. It is currently on display as part of the Egyptian tomb installation at the Metropolitan Museum of Art in New York City.

River, flowing through regions inhabited by spirits of the dead. Similar allusions to the Milky Way as a divine river are found in the cosmologies of the ancient Chinese and Arabic cultures. These and other cultures also saw the Milky Way as a vast bridge or roadway leading to immortality.

To the Hellenic Greeks, the origin of the Milky Way was but one part of their melodramatic myths starring pantheons of quarrelsome and mischievous gods. According to one legend, baby Heracles (Hercules) was brought by Hermes to suckle at the breast of Zeus's slumbering wife, Hera, and so gain immortality. Upon waking and learning that she was nursing not her own child but instead the son of Zeus and a mortal concubine, Hera pushed Heracles away, her breast milk spraying into the heavens. This convoluted tale of origin has been painted by several artists, including Rubens and Tintoretto. The familiar appellation "Milky Way" may have emerged from this legend, beginning with the Greek "Galaxias Kuklos," or "Milky Circle," and later becoming the Latin "Via Lactea," or "Milky Way." From these antecedents we have also inherited the word "galaxy," whose root simply means "milk."

In the Southern Hemisphere during the months of May through August, the brightest parts of the Milky Way appear highest in the sky. The resulting luminescence, when witnessed in a dark sky, is one of the most astonishing of naked-eye sights. The irregular structure of this diaphanous apparition prompted the Desana Indians of the Amazon rainforest to think of two intertwining snakes—one a male rainbow boa, the other a female anaconda. The Quechua Indians in the Andean foothills saw animal figures in the dark features that curdle the band of milky light (see Figure 6.2). These depictions are among several ancient descriptions of the structure inherent to the Milky Way.

Galileo Galilei was the first to resolve the Milky Way into myriads of stars, his observations in 1610 having been made only a year or so after the invention of the refracting telescope. In his own words, "we are at last freed from wordy debates about it. The Galaxy is, in fact, nothing but a congeries of innumerable stars grouped together in clusters. Upon whatever part of it the telescope is directed, a vast crowd of stars is immediately presented to view."

Using a larger reflecting telescope, Sir William Herschel and his sister Caroline surveyed the Milky Way in the 1780s, "gauging" the numbers of stars as a function of location in the sky. By assuming that all stars were of equal intrinsic brightness, that they were distributed with uniform density, and that there was nothing to obscure the starlight, the Herschels derived a map of the Milky Way as viewed from beyond it. From their assumptions, they reckoned that those regions showing the most stars must extend the

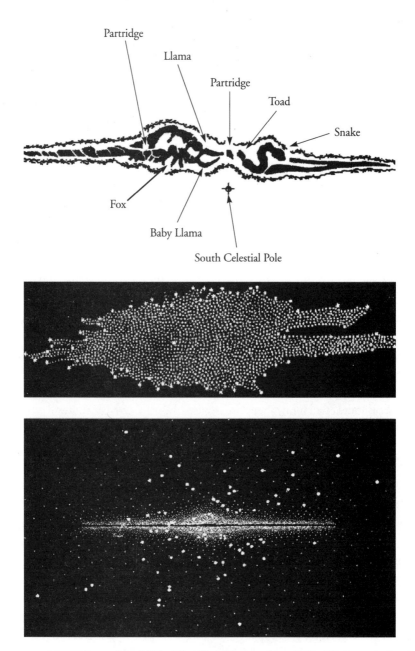

FIGURE 6.2 Visions of the Milky Way. Top: The southern Milky Way, as seen by the Quechua Indians of Peru. The Quecha interpreted the dark regions as mythic animals. Today, the dark regions are recognized as silhouetting clouds of dust in our galaxy. Middle: The Milky Way as viewed from outside, according to the "star gauging" survey carried out by William and Caroline Herschel in the 1780s. The Sun's location is denoted by the slightly off-center dot. Bottom: A "modern" view of the Milky Way, following Harlow Shapley's discovery of the Sun's offset position with respect to the distribution of globular clusters.

greatest distance from us. The resulting map is shown in Figure 6.2. It turns out that none of their assumptions was correct. Nevertheless, their map did get one thing right—the Milky Way is a flattened system. In retrospect, both the flatness of the Milky Way and its concentration toward the constellation of Sagittarius are apparent by just looking at the night sky during the months of May through August—when Sagittarius is highest in the sky.

This naked-eye impression was finally confirmed by Harlow Shapley in 1917, when he estimated that our Galaxy's system of globular clusters was centered on a point roughly 50,000 light-years away in the direction of Sagittarius. By reasoning that the 100 then-known globular clusters were tracing the overall gravitational field of the Milky Way, he concluded that we lived in a vast highly-flattened galaxy, roughly two-thirds of the way out from its center (see Figure 6.2).

Since Shapley's seminal discovery, we have become accustomed to our Sun's place as an off-center and minuscule part of a magnificent spiral galaxy. The numbers have changed, with the solar orbit lying some 28,000 light-years from the center, the stellar disk ranging out to about 40,000 light-years in radius, and the total extent of the globular cluster distribution being "only" 100,000 light-years in radius. We have also come to recognize that our Galaxy is but one of billions. It is worth remembering, however, that not too long ago the best astronomers were convinced that the Sun was at the center of all that was then known—despite what we would today regard as obvious clues to the contrary.

The Multi-Wavelength Milky Way

Today, our best visual impressions of the Milky Way reveal a panorama of bright and dark features on all conceivable scales (see Plate 6). There is so much to see that it is difficult to know where to begin. One highly recommended way is simply to lie down and slowly survey the Milky Way with a pair of good binoculars. This exercise provides one of the best introductions to the multitudes of star clusters, emission nebulae, and dust clouds that inhabit our home galaxy.

After becoming better acquainted with the visible Milky Way, you can then begin to appreciate the amazingly different vistas at nonvisible wavelengths. At near-infrared wavelengths, the effects of obscuration by dust are greatly diminished, so you can get a clear view of our stellar disk (see Plate 6). The disk itself is much thinner than that inferred from the visible Milky Way, in large part because the visible view is biased by nearby unobscured stars. A similar effect is evident in a forest, where your immediate view is dominated by the nearest trees and bushes. If these nearby objects were taken away, the view would be dominated by a much thinner band of greenery along the horizon. At near-infrared wavelengths, the more distant stars in the disk prevail, and being more distant, they appear closer to the midplane of the disk.

The near-infrared view also makes sensible the coordinate system of Galactic longitude and latitude. The Galactic longitude is set to zero in the constellation of Sagittarius—where the near-infrared (IR) emission is brightest. As seen in all-sky mappings (such as that in Plate 6), the longitude increases from the center toward the left, until it reaches 180° at the left edge of the map (at the boundary between the constellations of Auriga and Taurus). This same longitude is also mapped at the rightmost edge, with the longitude again increasing to the left, until it reaches 360° (= 0°) at the center. The Galactic latitude is measured from the midplane of near-infrared emission, with positive latitudes extending from the midplane up to +90° at the North Galactic Pole (in the constellation of Coma Berenices), and negative latitudes extending down to −90° at the South Galactic Pole (in the constellation of Sculptor).

At far-infrared wavelengths, the interstellar dust radiates away its star-warmed heat. Because the dust is entrained with the gas, the far-infrared view of the Milky Way provides an excellent roadmap of the gas and its concentrations (see Plate 7). Most of the action is found close to the Galactic midplane. But several nearby star-forming clouds can be seen at high Galactic latitudes. A similar blend of dominant midplane emission plus some high-latitude features is evident at radio wavelengths (see Plate 7). The radio continuum emission here is from electrons in the gas, wherever they may roam. Again, most of the high-latitude features correspond to nearby clouds of gas. The most prominent of these is the North Polar Spur—thought to be part of the Local Bubble that envelops the Solar Neighborhood with warm ionized gas.

Completely different views are obtained at shorter wavelengths and correspondingly higher photon energies. In X-rays, interacting binary stars punctuate the darkness, while in gamma rays the interaction between cosmic rays and the interstellar gas yields up an abundant harvest of these highest-energy photons. Only recently have we been able to map the Milky Way at these energies, and so our interpretations of the salient physical processes are not very secure. For example, some X-ray mappings indicate unusual goings-on in the general direction of the Galactic center but at high Galactic latitudes. Could these features indicate a huge outflow from the inner reaches of the Milky Way, or do they represent structures closer to us?

Denizens of the Milky Way

A complete census of the marvels to be found in the Milky Way is far from being realized. Nevertheless, our proximity to all that inhabits this great galaxy allows us to examine a few of these wonders in great detail. Let us begin with the Orion star-forming region.

Located behind most of the visible stars in the constellation of Orion lies the nearest of the Milky Way's giant molecular clouds. The Orion Molecular Cloud is roughly 1,500 light-years away and contains almost a million Suns worth of molecular gas—mostly in the form of diatomic hydrogen (H_2). Millimeter-wave observations of the carbon monoxide (CO) emission from this cloud reveals a loopy filamentary structure (see Figure 6.3). The dust associated with the ponderous cloud can be mapped via its far-infrared emission (see Plate 8). The most intense emission indicates the presence of newborn hot stars, their profligate ultraviolet (UV) radiation warming the dust and so inducing the observed knots of far-infrared light.

In this one molecular cloud complex (Figure 6.3 and Plate 8), the variety of emitting structures is almost too great to fathom. But several optical counterparts provide a basis for understanding what we see. From the top, a circular shell-like feature of far-infrared emission can be perceived. This shell surrounds the hot O8-type double star Lambda Orionis (Meissa), more familiar to the amateur astronomer as the head of Orion. Ultraviolet photons from this torrid star couple have been absorbed by the surrounding gas, ionizing and heating it. The heated gas, in turn, has expanded into the cooler non-ionized medium. The resulting pile-up of gas and dust is visible as the far-infrared shell, while the ionized gas interior to it can be traced via the ruby-red light that is emitted by the recombining hydrogen atoms. Here, we are witness to an almost perfect Stromgren Sphere, named after Bengt Stromgren, who first worked out the balance of photo-ionizations (ionizations by energetic photons) and recombinations back to the atomic state that governs the radiative energetics of H II regions.

Several other H II regions can be seen in the Orion field (Plate 8). Directly below Lambda Orionis, in a region adjoining the easternmost star in Orion's belt (Zeta Orionis, or Alnitak the Girdle), a luminous knot of warm dust indicates the position of the Horsehead Nebula. As shown in Plate 9, the Horsehead itself is a protrusion of dust and molecular gas that has yet to be evaporated and ionized by the UV radiation from the nearby O9 star Sigma Orionis. The surface of this molecular cloud radiates profusely in the mid-infrared, where spectral emission features due to polycyclic aromatic hydrocarbon (PAH) molecules abound (see Plate 9).

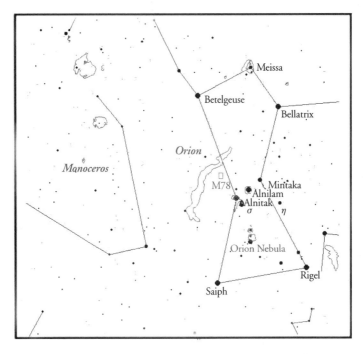

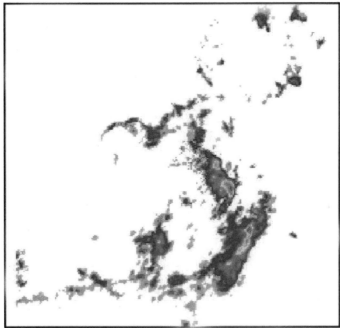

FIGURE 6.3 The Orion constellation, outlined at the top, as mapped at the 2.6-millimeter radio wavelength of carbon dioxide (CO). The CO mapping (bottom) reveals a giant molecular cloud complex that spans nearly the entire constellation of Orion and contains close to a million Suns worth of gas, mostly in the form of diatomic hydrogen (H_2).

At optical wavelengths, the dusty Horsehead appears in silhouette against diffuse emission from ionized gas in the background (see Plate 9). At the point where the horse's "jaw" meets its "neck," a tiny star can be seen. The winds from this newborn star may be responsible for having originally split the jaw from the neck, and may continue to excavate the neighboring nebulosity for many more millennia. Meanwhile, the ultraviolet torrent from Sigma Orionis is expected to irradiate the horse's head and flanks for another few million years, after which the star is likely to explode. The resulting salvo of shock waves should tear asunder whatever is left of the nebula.

Continuing down from the region of the Horsehead Nebula and into the middle of Orion's "sword," we encounter the most intense far-infrared source in the Orion field. The optical counterpart is well known to amateur astronomers as the great Orion Nebula (see Plate 10). Listed as number 42 in Charles Messier's 1781 catalogue of fuzzy objects in the sky, the Orion Nebula is the closest region to us where burgeoning high-mass star formation prevails. It happens to be on the near surface of the Orion Molecular Cloud and so is visible to us. Underlying all the dazzling nebulosity is a cluster of some 1,000 stars, spanning about 10 light-years, and thought to be no more than 2–3 million years old. The most massive and luminous stars are concentrated in a tight asterism of four stars known as the Trapezium, and along a nebular "bar" that delineates one edge of the inner nebula (see Plate 10). These hot O-type stars, and in particular the O6-type star Theta Orionis C in the Trapezium, are most responsible for powering the nebular emission that is evident at optical, infrared, and radio wavelengths.

The intense power of Theta Orionis C can be surmised from the effect it is having on lesser stars in its immediate vicinity (see Plate 10). Here, we see cometary structures emanating from these smaller stars away from their domineering neighbor. It is possible that the comet-like tails have been stripped off the smaller stars by the superwind of Theta Orionis C. Theorists, however, argue that if this is so these stars would be surrounded by bow shocks, and so would never directly feel the superwind. Instead, the tails may indicate that the O star's intense ultraviolet radiation is ablating the stars through some sort of photoevaporative process. Alternatively, wind and light shadows cast by the stars themselves may have led to enhanced concentrations of nebular gas in the shadows, forming the tails.

Despite all the violent activity associated with the hot stars in the Orion Nebula, more mundane activity proceeds apace. Indeed, the inner nebula is currently host to several protostellar systems, complete with protoplanetary disks. These disks (dubbed proplyds) can be seen in silhouette against the light of the background nebulosity (see Plate 10). The fact that protostellar and protoplanetary processes continue in the Orion Nebula tells us that star and planet formation is an incredibly robust process. Although we may never know for certain, it is entirely possible that the Sun and Solar System were once formed within an H II region such as the Orion Nebula.

Approximately 4,000 giant molecular clouds similar to that in Orion have been found lumbering about the Milky Way galaxy. These behemoths contain more than half of the molecular gas in the Milky Way, and so it is not surprising that most of the known optical and radio H II regions are associated with them. As the embedded newborn star clusters evolve, their most massive stars die off in titanic supernova explosions, and their surrounding cocoons of gas eventually dissipate into the interstellar medium (see Plate 11).

What remain are fully exposed clusters of intermediate- and low-mass stars. Nearby examples include (in order of increasing age) the Orion Nebula cluster (2 million years, or Myr), the double cluster (h and chi) in Perseus (10 Myr), the Pleiades (M45) in Taurus (100 Myr), and the Beehive (M44) in Cancer (1,000 Myr) (see Figure 6.4). All of these clusters are known as open, or galactic, clusters. They typically contain no more than 1,000 stars in fairly loose arrangements measuring less than 100 light-years across. The tidal stresses from the Milky Way itself, along with occasional encounters with giant molecular clouds, are thought to disrupt the orbits of stars in these clusters, thus accounting for the lack of any open clusters older than 7 billion years. Approximately 1,000 such clusters have been observed in the Milky Way.

The looser associations of O- and B-type stars are even more vulnerable to the vicissitudes of their Galactic environment. Extending for 100–1,000 light-years, these ephemeral constructs of young hot stars are thought to represent the stellar residue of giant molecular cloud complexes that have recently dissipated into the interstellar medium. In some OB associations, such as the Orion OB1 association, subgroups are arrayed according to age, which indicates that some sort of sequential mode of star formation has occurred. About 100 associations are known in the Milky Way, their rarity being a direct consequence of their brief lives.

Farther away, filling the halo of the Milky Way, the globular clusters exhibit completely different characteristics. Prominent examples include the Great Cluster in Hercules (M13) in the Northern Hemisphere, and 47 Tucanae (NGC 104) and Omega Centauri (NGC 5139) in the Southern Hemisphere (see Figure 6.4). These objects contain as much as a million stars in tight balls measuring 100 light-years or so across. Their corresponding color-magnitude diagrams indicate an absence of main-sequence stars

FIGURE 6.4 Comparison of the open star clusters h and chi in Perseus (top) with the globular cluster M13 in Hercules (bottom). Each view measures approximately 150 light-years on a side, and so highlights the true differences between these two types of clusters.

more luminous than the Sun. The complete lack of luminous, short-lived stars indicates that these stellar snarls must be at least 10 billion years old. The latest and most precise distances provided by the Hipparcos astrometric satellite yield absolute luminosities of the individual stars that indicate ages between 12 and 14 billion years.

These clusters represent our most ancient relics from a time when the Milky Way was more like a sphere than a pancake. This was a time characterized by stars and star clusters in highly elliptical and randomly oriented orbits, as the proto-Galaxy collapsed under its own weight. Elements heavier than helium were virtually absent then, which means that no rocky planets could have formed with the stars. Today, these clusters are dim leftovers of what they once were—when they still harbored massive stars, each star as luminous as the remaining 100,000 stars combined.

Constructing Our Galaxy: The Stars

Like curious caterpillars in a cosmic forest, we are sorely challenged in our efforts to learn the overall lay of the Galaxy. We cannot help taking some cues from nearby galaxies, which can be studied in considerable detail. Indeed, several spiral galaxies in edge-on configurations appear remarkably like our basic visual impression of the Milky Way (see Figure 3.6). To work out the actual architecture of our Galaxy, however, it is necessary to derive distances to all that we see or, conversely, to generate galactic models that predict what we should see from our particular vantage point, and then compare these predictions with our basic observations. Both methods have been pursued, with some successes worth reporting. Yet even today, considerable disagreement remains over the basic specifications of the Milky Way.

The Spheroidal Component

The most straightforward component to fathom is the spheroid, or halo, that envelops the disk of our Galaxy. There, the objects are not subject to the amounts of obscuration that prevail in the disk. We are also aided by the globular clusters, whose individual stars can be observed and then collectively analyzed in terms of distance, age, and metal abundance. What we find is a vast, tenuous, and metal-poor stellar component that extends out to a mean radius of 100,000 light-years. If the most distant globular clusters and satellite dwarf elliptical galaxies are included as part of the halo, then the total extent increases to 300,000 light-years—encompassing the Large and Small Magellanic Clouds, some 180,000 light-years away (see Chapter 7).

As previously noted, the centroid of the spheroidal component is located in the direction of the constellation Sagittarius. Studies of RR Lyrae variable stars, pulsating yellow giants that inhabit the halo, have shown that they share a near-constant average luminosity of about 50 times that of the Sun. Because of their well-defined luminosities, RR Lyraes have served as "standard candles" for reckoning distances throughout the halo. The resulting halo distribution is centered 28,000 light-years away. The space density of RR Lyraes is best fit by an inverse-cube ($1/r^3$) decline with radius (r), which indicates a spheroidal component with a very steep falloff. As observed from another galaxy, the halo would appear as a two-dimensional object whose brightness distribution is a summation of all the stars along each line of sight. The resulting numbers of stars observed at each radius follow a less precipitous inverse-square decline. Such falloffs are similar to those seen in elliptical galaxies and the bulges of other spiral galaxies.

The actual shape of the spheroidal component is rather uncertain. If it follows the overall gravitational field of the Galaxy, the disk's gravitating contribution should cause it to be somewhat squished perpendicular to the disk. The distribution of globular clusters does show some flattening within 10,000 light-years of the Galactic center, but farther out it appears for the most part spherical. Models incorporating large amounts of dark matter in the halo yield similarly spherical configurations.

By counting the proportion of RR Lyraes in globular clusters and their total number in the rest of the halo, it is possible to estimate the total luminosity and mass of stars in the halo. The visual luminosity turns out to be very small, roughly 2 billion Suns worth, or about one-sixth the visual luminosity of the disk. The stellar mass in the halo is estimated at roughly 24 billion solar masses, or about half the disk mass. Dynamical considerations suggest that the halo is hiding more than 10 times this amount in the form of some sort of gravitating dark matter. Whatever that matter is, it must be incredibly old to match the 10–15-billion-year ages characteristic of the halo stars.

The spheroidal component nucleates into what is known as the bulge component, a dense stellar conglomeration measuring roughly 10,000 light-years across. Although much of the central bulge is hidden from our view by intervening clouds of dust, there are a few relatively dust-free "windows" through which we can discern red giants and other bright stars belonging to the bulge. One of the most famous is Baade's window, located in the so-called Great Sagittarius Star Cloud. Intensive studies of this region have revealed a unique blend of stars that defies easy interpretation.

In contrast to the metal-poor conditions in most of the halo, the bulge contains stars of both low and high metallicity—with a mean metallicity that is twice that found near the Sun. There also seems to be a wider range of stellar ages, although the most ancient stars still dominate. Apparently, the bulge has a stellar history as old as the rest of the halo, but includes some more recent chapters featuring replenishment of its gaseous stores followed by renewed star formation.

The metal-rich stars are especially intriguing, for they imply an early and intense epoch of starbirth and stardeath, whose ashes have enriched the later generations of stars that we observe today. But even this interpretation may be an oversimplification. Like a teeming metropolis, the bulge of our Galaxy contains a bewildering variety of stellar types, many of which have yet to be accounted for. Some no doubt belong to the halo, while others may represent the innermost parts of the Galaxy's disk.

The Disk Component

The disk of our Galaxy is most familiar to us, for that is what we see as the Milky Way. Much of what we know about the disk component has been obtained from stars within about 5,000 light-years of the Sun, or less than 1 percent of the total expanse. Although this tiny sampling of the disk may be unrepresentative, we don't have much of a choice in the matter. Dusty clouds obscure the visible light from most of the stars that lie beyond a radius of 5,000 light-years or so.

By counting the numbers of stars of differing type, distance, and direction, astronomers have pieced together some remarkable findings. The first major finding is that the dim, low-mass stars predominate over the bright massive stars. It really does seem that the "meek" have inherited the Galaxy. The degree of this predominance has been quantified in what is known as the local stellar luminosity function, whereby the number of stars is enumerated as a function of luminosity or absolute magnitude (see Figure 3.4). Because a star's luminosity is related to its mass and burning lifetime, astronomers can use the measured luminosity function to constrain the distribution of stellar masses as well as the overall star-formation history in our part of the Galaxy. For example, the small bump at an absolute magnitude of $M_V \cong 5.5$ mags (and corresponding mass of about 0.8 Suns) probably indicates that a burst of star formation occurred approximately 12 billion years ago—roughly corresponding to the estimated mean age of the stellar disk. Without such a burst, the luminosity function would have shown a continuous decline at greater luminosities, in keeping with a stellar mass distribution that is biased against high-mass stars.

Another major finding pertains to the thickness of the stellar disk and its dependence on stellar type. The hotter, brighter, and more massive stars occupy a much thinner disk than the cooler, dimmer, and less massive stars. Again, the relationship between mass and total burning lifetime provides the key to interpreting this result. Because high-mass stars are characteristically younger, their much thinner distribution probably has something to do with their youth.

One explanation is that stars are created in clouds that are susceptible to mutual collisions. As the clouds radiate away the energy of the collisions, their relative motion is spent, and they end up settling down very close to the Galactic midplane. Once formed in the clouds, the stars are exposed to gravitational perturbations by other clouds and star clusters in the galaxy, with the exposure and its effects accumulating over time. Ultimately, the stars are "heated up" by the perturbations, and their increasing noncircular velocities cause them to stray farther from the Galactic midplane.

An alternative explanation posits that the more ancient stars were formed in thicker disks, which these stars now trace. This scenario borrows from the situation of the spheroidal component stars, which were probably formed during the proto-Galaxy's initial collapse. After having spawned the thick stellar disk, the gaseous disk would have slowly settled down over time into the very thin disk that today hosts star-forming clouds and young stars. Such an interpretation, however, fails to explain the dominance of circular co-planar orbits—even for the older stars in the thick disk. It also leaves open the question of how a thicker disk of more tenuous gas could have congealed into star-forming clouds. That is why many astronomers favor the scenario in which the proto-Galaxy initially condensed into a spheroidal stellar configuration along with a thin gaseous disk, whose resulting stellar populations have fluffed up over time.

Thanks to the Cosmic Background Explorer (COBE) mission, we now have maps of the integrated starlight from our Galaxy (see Plate 6). At wavelengths of 1 to 5 microns, the light from red giant stars is readily detected, while obscuration by dust is minimized. Recent modeling of these light distributions has led to some far-reaching conclusions. As shown in Figure 6.5, our Galaxy appears to harbor a pronounced central bar. The bar's long axis is directed almost toward us, the deviation being only about 13 degrees. The most obvious effect of the bar as seen in the COBE images is a slightly lopsided central "bulge," the consequence of the bar's being severely foreshortened as viewed by us.

According to the best-fitting model, the COBE near-infrared maps indicate that the bar in our Galaxy's disk incorporates almost all of what was previously regarded as the bulge component. Such an astonishing result will require further testing. But the existence of the bar itself gains further support from the noncircular orbits of gas evident in the inner Galaxy, and from the ring of molecular gas that appears to encircle the putative bar (see the next section). Such ring-bar configurations are observed in other nearby spiral galaxies, and are thought to be caused by orbital resonances between the bar and the gas in the disk (see Chapter 9).

Beyond the bar, the stellar disk thins out pretty much as expected. But toward its extremities it has one last hurrah—warping like the brim of a fedora. Warped disks have been seen in other spiral galaxies, especially in the more extensive atomic hydrogen distributions of these galaxies. The dynamics be-

FIGURE 6.5   Face-on (top) and edge-on (bottom) renditions of the Galaxy's old stellar disk, without the spiral and other features associated with recent star formation. This model is based on analysis of the near-infrared emission as mapped by the COBE satellite (see also Plate 6). The Sun is depicted as a tiny dot below the almost vertical bar in the face-on view. As modeled, the major axis of the central bar is oriented within 15 degrees of the imaginary line connecting the Sun with the nucleus. The edge-on view shows a slight warping in the outer stellar disk.

hind these rather stylish structures are usually linked to one or more gravitating perturbers. In the case of the Milky Way, likely candidates are the Large and Small Magellanic Clouds, whose orbits are not co-planar with the disk of the Milky Way. The tidal tugging from these companion galaxies produces extra-planar torques on the stars and gas clouds that are orbiting in the outer disk of the Milky Way. These torques ultimately yield tilted orbits, whose collective manifestation is the warping that is observed.

Constructing Our Galaxy: The Gas

The dust that obscures so much of our visible view keeps company with a bubbling froth of gaseous phases (see Plate 12). Throughout the disk and

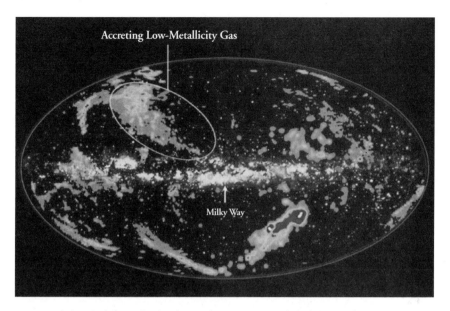

FIGURE 6.6 High-latitude cloud complexes, as mapped in the 21-centimeter emission line of atomic hydrogen. A drawing of the visible Milky Way, included for reference, fills the midplane of this all-sky mapping. Complementary observations with optical telescopes have shown that the high-latitude clouds are not nearby but lie 10,000 to 40,000 light-years away in the halo of our Galaxy. The trajectories of these clouds indicate that at least some of them are raining low-metallicity gas onto the Galaxy's disk, thus replenishing the disk's store of primordial gas. The Large and Small Magellanic Clouds dominate the lower right quadrant, with the enveloping Magellanic Stream of atomic hydrogen sweeping over the Galactic South Pole to the lower left.

halo, most of the volume is occupied by the hottest phase, the so-called coronal gas. This extremely tenuous medium, with densities on the order of 1–10 particles per liter (1,000 cubic centimeters), has temperatures of 10^5 to 10^7 degrees Kelvin. The resulting pressures are sufficient to maintain an equilibrium of sorts with the much cooler and denser phases found in the disk. Indeed, the starbursting pyrotechnics in the disk may be responsible for generating and injecting this hot plasma into the halo. In turn, the slowly cooling coronal gas may ultimately feed the disk, with the combined circulation resembling that of a galaxy-wide fountain.

Coursing through the coronal gas are gigantic cloud complexes of much cooler neutral hydrogen gas (see Figure 6.6). Approximately a dozen such cloud complexes are known, each one measuring about 1,000 light-years across and containing some 10^5–10^7 solar masses of gas. These colossi can be distinguished from the more common clouds of neutral hydrogen in the disk by virtue of their much greater deviations from the Galactic midplane and by their anomalous Doppler velocities. Upon their discovery in the 1960s, they were designated high-velocity clouds. But astronomers have since realized that the hydrogen in the rotating disk is actually the high-velocity component, while at least some of the high-latitude clouds show very little rotation and are instead plunging toward the disk on highly elliptical trajectories.

The most famous of the high-velocity cloud complexes spans almost 90 degrees of the southern sky. This so-called Magellanic Stream begins as a relatively concentrated envelope of gas around the Large and Small Magellanic Clouds, slowly thinning out as it follows a curved path across the sky. In 1999, a similar spur of gas was identified that links the Clouds to the Milky Way. There, the gravitationally dominant Milky Way is probably stripping and consuming gas from these companion irregular galaxies (see Chapter 7).

At optical wavelengths, the high-latitude clouds absorb the light from distant galaxies and quasars. The detailed spectrum of absorption reveals a dearth of elements heavier than hydrogen and helium. Therefore, at least some of the high-latitude clouds are primordial rather than the cooling ejecta from starburst activity in the disk of the Milky Way or in the Magellanic Clouds. The fact that some of these clouds appear to be accreting onto the Galactic disk provides present-day support for the idea that galaxy disks were built up via the accumulation of satellite clouds.

Not all of the high-latitude clouds are part of our Galaxy, however. Of the few clouds whose distances have been measured, a subset was identified in 1999 with distances placing the clouds in this subset far beyond the Milky Way galaxy. Roaming among members of the Local Group, these starless clouds may represent primordial, unborn galaxies that have lain dormant since their emergence some 10–15 billion years ago.

Within the disk, the interplay between the interstellar medium and the ensuing starbirth activity has led to a complex architecture of gaseous phases. The densest and coolest phase consists of molecular clouds. Each molecular cloud consists primarily of diatomic hydrogen (H_2) along with trace amounts of carbon monoxide (CO) and more complex molecules. Such a cloud typically measures about 10 light-years across and, with a density of a million or more molecules per liter, contains roughly 10,000 solar masses of the molecular gas. The rarer giant molecular clouds, such as those in Orion, span about 100 light-years and contain more like 10^5–10^6 solar masses each. Despite their rarity, the giant molecular clouds are so massive that they constitute the bulk of the molecular phase. By far, they are the largest, most ponderous single objects in the disk.

CO emission-line surveys of the Milky Way, carried out in the 1980s with millimeter-wave radio telescopes, clearly showed that the molecular gas in

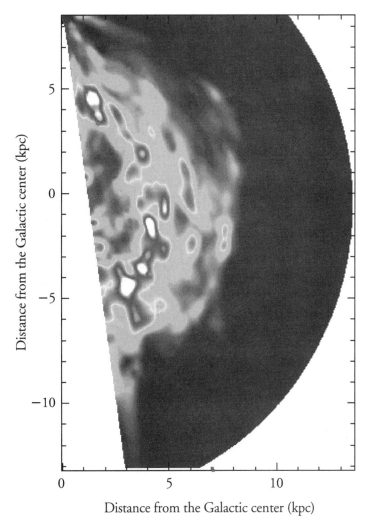

FIGURE 6.7   This face-on rendering of molecular hydrogen (H$_2$) in the Milky Way is based on a kinematic analysis of the CO spectral-line emission near the Galactic mid-plane. The Sun is located at the upper left corner of the map. The molecular gas appears to have a ringlike or very tightly wound spiral structure interior to the Sun's orbit.

the disk is confined to an inner ring (see Figure 6.7). The molecular ring is located about 16,300 light-years (or 5,000 parsecs) from the center of the Galaxy, just beyond the extremities of the stellar bar discussed earlier. Associated with the ring is a profusion of radio-continuum emission, whose origin is most likely gas that has been ionized by newborn hot stars. Observers on another galaxy would find our molecular ring aglow with the ultraviolet light of O- and B-type stars along with the fluorescent-red light of recombining hydrogen gas. In other words, they would see a barred spiral galaxy with a starbursting ring encircling the bar.

The neutral hydrogen gas shows a completely different distribution. Instead of a ring, the radio-emitting atomic hydrogen (H I) is configured like a compact disk—complete with a hole in the middle. The gas itself is organized into less distinct clouds with masses similar to those of the giant molecular clouds. Some astronomers argue that many of the molecular clouds have atomic envelopes, a configuration that which further complicates the multi-phase architecture.

Comparisons between the 21-centimeter radio emission from H I and the far-infrared continuum emission from dust indicate strong spatial similarities. From the higher-resolution, far-infrared images, we can therefore surmise that the H I has been shaped into a "froth" of filaments, loops, and shells (see Plate 12). Spatial analysis of this froth reveals fractal behavior, in that there is a statistically smooth and quantifiable hierarchy of structure on all measured spatial scales. Somehow, the ambient mix of self-gravity, differential shearing, radiation pressure, thermal instability, magnetic fields, stellar winds, supernova blast waves, and other agents of turbulence has led to this marvelously baroque arrangement of nebular matter.

Beyond the solar orbit, the neutral hydrogen disk flares to greater heights above and below the galactic midplane. This flaring is caused by the decrease in stellar density and corresponding diminution in gravitational binding of the gas to the midplane. As the gaseous density plummets, so too do the number of discrete molecular clouds and star-forming regions. Beyond a radius of 40,000 light-years, the disk of the Galaxy becomes a barren place, traced only by its outlying atomic gas out to another 20,000 light-years or so.

The last major gaseous phase associated with the disk was discovered in the 1980s. This diffuse-ionized gas component is a more tenuous version of the gas that characterizes H II regions. It was found through sensitive wide-field surveys of the spectral-line emission from diffuse hydrogen, nitrogen, and sulfur ions in the sky. The structure of this component is currently unknown, but new surveys are expected to yield detailed and comprehensive maps of this important interstellar phase.

Unlike the molecular component, which has an overall thickness of about 1,000 light-years, the diffuse-ionized gas strays more than 10,000 light-years from the Galactic midplane. Such enormous extent, combined with a density of about 1,000 atoms per liter along with a much greater fractional filling of the Galactic volume, results in an overall mass that is comparable to those of the molecular and atomic phases.

The source of ionization is thought to be the ultraviolet radiation from O-type stars in the disk, although it is possible that interstellar shock waves may be playing some additional role in exciting the gas. From the line emission, we can infer that the diffuse-ionized gas has been enriched with heavier elements, like the rest of the gas in the disk. Therefore, this diffuse plasma probably represents the final vanguard of our gaseous disk as it melds with the halo, rather than some precipitate from the primordial high-latitude clouds.

The molecular, atomic, and ionized gas components each contribute about 5–10 billion solar masses to the total content of the Milky Way. All told, they constitute roughly 20 billion solar masses—about 20 percent of the Galaxy's total luminous matter, and about 5 percent of the total dynamical mass (including the so-called dark matter). In the past, the gaseous contribution was much higher, with fewer generations of stars having locked up the original reserves. Today, we find our Galaxy at the tail end of its transformation from a gaseous blob into a predominantly stellar system.

Galactic Motion

When Galileo Galilei was compelled by the papal prelates to recant his claim that the Earth orbits around the Sun (rather than vice versa), he was heard to mutter, " . . . yet it moves." Since that time, civilization has come to recognize the profundity of Galileo's words. Today, we are witness to a swirling hierarchy of cosmic motions. To examine the motions of our Galaxy, let us begin with the Earth and work our way out.

Our home planet spins about its axis every 23 hours, 56 minutes, and 4.1 seconds. This whirling translates to a surface velocity at the equator of 1,677 kilometers per hour (0.47 kilometers per second), roughly twice that of a jet airliner, diminishing to jet speed near the Arctic and Antarctic Circles. The Earth orbits the Sun every 365.26 days. Its mean velocity of 29.9 kilometers per second during its nearly circular elliptical orbit is about 64 times faster than its equatorial spinning.

The Sun has a peculiar motion relative to its neighboring stars, which on average amounts to a velocity of 19.5 kilometers per second toward the constellation Hercules (Galactic longitude $l = 56°$, Galactic latitude $b = 23°$). This trajectory can be ascertained by surveying the motions of stars toward and away from us (inferred from their spectral Doppler shifts) and their motions across the sky (measured in units of arcseconds per year, one arcsecond being 1/3,600 of a degree). In the direction of Hercules, the line-of-sight motions are most pronounced and headed our way. That is because the Sun is actually approaching these stars at a mean velocity of 19.5 kilometers per second. In the same direction, the stellar motions across the sky are minimal, and that indicates minimal cross-wise motion of the Sun relative to these stars.

The Sun's peculiar motion directs it slightly out of the Galactic plane (by 23 degrees). As it orbits around the Galaxy, the Sun bobs up and down through the Galactic midplane—like a fiery porpoise in a circulating school of 100 billion other blazing porpoises. From the estimated density of matter in the Galactic disk near the Sun, the restoring force against vertical motions can be derived, and from that, the period of vertical bobbing. This bobbing period turns out to be about 62 million years—remarkably similar to twice the average timespan between extinction events on Earth.

Some astronomers have speculated that the Sun's periodic midplane crossings result in close encounters with giant molecular clouds. The disruptive gravitational influence of these clouds on the outer Solar System could have caused showers of comets to fall into the inner Solar System—thus wreaking havoc on Earth every 30 million years or so. But the amplitude of the bobbing (about 100 light-years) is no larger than the typical sizes of the perturbing clouds, and so there should not be much of a periodic effect.

Other astronomers have proposed that the even rarer mass extinctions on Earth have resulted from the Sun's less frequent traversals through spiral arms, where the giant molecular clouds are most concentrated. Such a scenario is supported by the apparent clumping of interstellar matter along spiral arms, and the generally good fit between past arm traversals and mass-extinction events. This theory predicts the next mass extinction will occur in another 60 million years or so, when the Sun traverses the so-called Perseus Arm (see the next section).

The Sun's peculiar motion also directs it toward the Galactic center (by 34 degrees). This noncircular motion is thought to be part of a general oscillation toward and away from the Galactic center. Compared to the vertical bobbing, the radial oscillation is a more sedate affair, taking about 169 million years (two-thirds its orbital period) to go through one cycle. As viewed from above the Galaxy, the Sun's radial excursions add to its rotational motion, producing a rosette-like orbital pattern.

The Sun's co-moving neighborhood of stars is known as the Local Standard of Rest (LSR). It is defined as the reference frame, instantaneously centered on the Sun, which moves in a circular orbit about the Galactic center at the circular speed appropriate to its position in the Galaxy. The LSR extends beyond the Sun for several hundred light-years, and encompasses several thousands of stars.

The LSR's motion about the Galaxy is determined by looking at the stars just beyond the LSR and investigating their rate of shearing. The shearing, also known as differential rotation, is manifested by the periodic variation in stellar line-of-sight velocities, or radial velocities, as a function of Galactic longitude. If the Galaxy were like a rotating compact disk, there would be no shearing. Indeed, all of the stars would be fixed relative to the Sun—with no discernible radial velocity. The significant radial velocities that are observed and the fact that they oscillate over 360 degrees of Galactic longitude indicates, however, that shearing does exist. Stars interior to the Sun sweep past those exterior to the Sun by about 9 kilometers per second per 1,000 light-years of Galactocentric distance.

The Dutch astronomer Jan Oort was the first to incorporate this differential rotation into a theoretical framework for determining the overall rotation of the Galaxy. From the formulations that he derived in 1927, the observed differential rotation yields an orbital period for the LSR of 240 million years. If we adopt the currently favored Galactocentric distance of 8.5 kiloparsecs (27,700 light-years), we can convert this orbital period into a circular speed of 220 kilometers per second—roughly 7 times faster than the Earth in its orbit around the Sun, and about 10,000 times faster than a jet airliner. At this speed, the Sun has made about 19 laps around the Galaxy during its 4.6-billion-year lifetime. In the process, its peculiar motion of 20 kilometers per second has caused it to drift away from whatever stellar kin it may have had at its birth. We find our home star to be a stray among strays.

Oort's kinematic formulations also enable us to determine the Galaxy's rotation far from the Sun. Measurements of the 2.6-millimeter line emission from interstellar CO reveal a distribution of radial velocities that broadly sweeps between −120 and +120 kilometers per second over 360 degrees of Galactic longitude. Here, the velocity mapping refers to molecular gas well away from the Sun, and so traces the overall rotation of the Galaxy. Through trigonometric reasoning, the observed distribution of radial velocities at each longitude can be translated into a solution for the circular velocity as a function of Galactocentric radius. The resulting rotation curve indicates a Galaxy with a fairly constant rotation velocity of about 200 kilometers per second out to twice the radius of the solar orbit (see Figure 4.3).

We are faced with rotational motions that far exceed what can be gravitationally constrained by the luminous matter in the Galaxy. From the stellar and gaseous distributions in the spheroid, bulge, and disk, one would estimate something like 100 billion solar masses of gravitating matter. Yet the kinematics of the gas in the disk indicates a gravitating mass more like 400 billion Suns. Even greater estimates are obtained by modeling the observed motions of star clusters and clouds in the halo—where most of the so-called dark matter is thought to reside. These dynamical investigations all indicate that at least 75 percent of the Galaxy's mass remains undetected and—so far—unexplained (see Chapter 4).

What Spiral Is This?

In less than 100 years, our Galactic world-view has transmogrified from a Sun-centered ellipsoid of stars measuring 50,000 light-years across to a vaster, bulge-centered spiral system of stars, gas, and dust embedded within a tenuous halo of ancient star clusters and gravitating dark matter. Comparisons with other nearby disk galaxies have fortified our supposition that we live in a spiral galaxy. Delineating the actual spiral structure of our Milky Way has proved to be a daunting task, however.

The best current evidence for spiral structure comes from the spatial distribution of massive short-lived stars within 10,000 light-years of the Sun. These stars include O- and B-type main-sequence stars as well as F- and G-type Cepheid variable supergiant stars, whose high luminosities make them visible at great distances. (We will discuss the special properties of Cepheid variables in the context of the Magellanic Clouds in Chapter 7.) Observations of other spiral galaxies invariably show these stars tracing out the spiral arms. Spectroscopic analysis of these stars indicates what kinds of stars they are, how intrinsically bright they should be, and how much their light has been obscured. Such analysis, when combined with measurements of the stars' apparent brightnesses, yields fairly reliable estimations of their distances.

Often, the massive stars are part of young clusters. By plotting the colors and apparent magnitudes of the stars in these clusters, astronomers can fit the observed main sequences to those of nearby clusters with well-known distances and so derive accurate distances to the more distant clusters. This photometric method for determining distances is known as main-sequence fitting.

The resulting spatial distribution of massive stars and young clusters is shown in Figure 6.8, where the Galactic center is situated well below the field in the direction of Sagittarius. Three parallel armlike features are clearly evident in the local disk, with the Sun, at the center of the diagram, appearing to lie near the inner edge of the middle feature. The farthest feature from

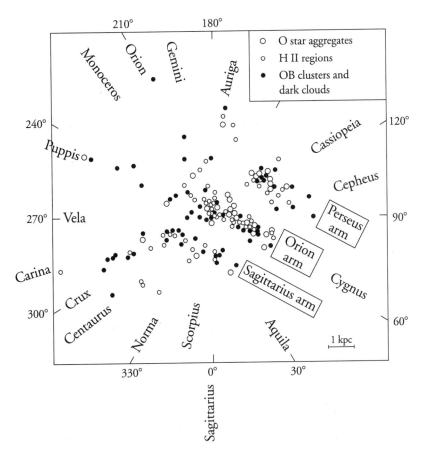

FIGURE 6.8 The spatial distribution of massive O- and B-type stars, H II regions powered by O-type stars, and star-forming dark clouds within 10,000 light-years of the Sun, which is situated at the center of the diagram. Three spiral-arm fragments can be discerned. This is the best evidence we have for spiral structure in the Galaxy.

the Galactic center, appearing in the direction of Perseus, is thought to delineate a small part of the so-called Perseus arm. The middle, "local" feature refers to the Orion arm, while the innermost feature traces the nearest segment of the Sagittarius arm.

The arms are separated by about 5,000 light-years, have pitch angles (as measured relative to Galactocentric circles drawn through them) of roughly 25 degrees, and trail with respect to the clockwise rotation of the Galaxy. A similarly derived map of old star clusters shows none of these armlike features. This disparity emphasizes the importance of short-lived massive stars in delineating spiral structure.

To investigate the Galaxy's overall spiral structure, it is necessary to work

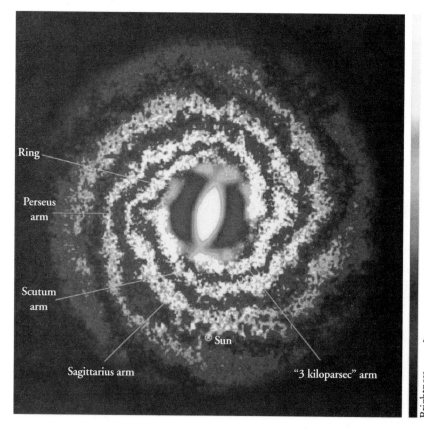

FIGURE 6.9   One version of the spiral structure in the disk of our Galaxy, including a prominent central bar and inner ring, based on analysis of near-infrared maps from the COBE satellite and—for the outer structures—radio mappings of giant H II regions. The four putative arms are seen to meld into the inner ring. Other candidate renditions show fewer arms with a greater number of windings.

with the gas. Emitting at radio wavelengths, the molecular, atomic, and ionized gas can be traced throughout the Galaxy without hindrance by the intervening dust. The Doppler-shifted radio spectral-line emission, in particular, provides unique information on the radial velocities of the gas. By adopting an independently derived rotation curve for the Galaxy, one can convert the radial velocities and longitudes of the prominent line emission into Galactocentric and line-of-sight distances—the latter yielding the spatial location of the emitting gas.

The combined results from the optical, infrared, and radio mappings are shown in Figure 6.9. Continuations of the optical spiral features are evident, along with additional arms. From this mapping, it appears that the Galaxy is a barred four-armed spiral with an inner ring. Alas, other mappings have yielded two- and three-armed solutions as well.

All of these mappings must be regarded with great caution. Because the line-of-sight distances are based on kinematic information, they are critically dependent on the assumption of circular orbits. Any noncircular trajectories in the gas will lead to erroneous attributions of distance and hence spurious mappings. And, like dogs chasing their tails, the spiral arms themselves produce just these sorts of noncircular streaming motions. As the orbiting gas approaches an arm, it is attracted by the arm's gravity to a greater Galacto-centric distance. It must then slow down in order to conserve angular momentum. After passing through the arm, the gas once again settles back to its interarm trajectory and higher orbital velocity. Such close and complex coupling between the gas structures and their kinematics leads to uncertainties of several thousand light-years in the derived distances—enough to blur any specific mapping of spiral structure.

Unlike a caterpillar in the forest, we have little hope of someday metamorphosing into some exquisite moth, taking wing, and seeing our galactic disk from high above it. Given these physical constraints, we can best guess that such a winged creature would behold a thin yellowish inner disk containing a central bar of old yellow-red stars, itself surrounded by an inner ring (or tightly wound two-armed spiral) of young blue stars that connects to multiple outer spiral arms with pitch angles of about 25 degrees. Reasonable analogues might be the ringed-barred spiral galaxies NGC 2336 and NGC 3992 (see Plate 2).

Nuclear Antics

If, in our imaginings, a giant moth could hover above the Milky Way, it would most likely head toward our Galaxy's brilliant nucleus. Unobscured by intervening clouds of dust, the pointlike nucleus would gleam like nothing else in the Galaxy. This fantastic scenario has a basis in empirical fact. The nuclei of other spiral galaxies are often much brighter than any other single sources in these systems. Moreover, multi-wavelength observations of our own nucleus have revealed a pointlike object embedded within a hotbed of stellar and gaseous emission.

The evidence for exciting goings-on in our nucleus is both extensive and perplexing. At optical wavelengths, the nucleus is obscured by so much dust that only one-trillionth of the light has a chance of getting through. At radio wavelengths, however, the dust has negligible influence and so the entire nuclear theater is exposed to view.

The pointlike source, known as Sagittarius A˙ (Sgr A˙), is observed in the radio continuum. Even with the most accurate radio mappings, Sgr A˙ remains marginally resolved. On the sky, it subtends less than 0.001 arcsecond. And at a distance of 27,700 light-years, that angular subtent translates to a linear size of only 1.1 light-hour, or about the distance between Earth and Saturn. Sgr A˙ is extremely close to the apparent dynamical center, a fact that supports the contention that it traces the true nucleus of the Galaxy.

Radio continuum mapping of the encompassing Sgr A West region has revealed a spiral "whirligig" of ionized gas that is centered on Sgr A˙ (see Plate 13). The three "arms" of this plasma structure have dimensions of 5–10 light-years. Studies of the radio line emission from these arms yield Doppler shifts consistent with inflow from the even larger torus of molecular gas that surrounds Sgr A˙. Perhaps we are seeing a low-energy version of the gas fueling that is thought to power the bizarre activity exhibited by Seyfert galaxies and quasars (see Chapter 11).

On much larger scales, the thermal radio continuum emission from the ionized gas in Sgr A West appears to connect with a gigantic ducktail of nonthermal (synchrotron) radio emission (see Plate 13). This feature extends away from the Galactic plane for about 100 light-years, tracing out disjointed magnetic fields in the surrounding interstellar medium. The overall impression is one of broad strokes with a paintbrush 100 light-years wide. Here, we see circumstantial evidence for some sort of magnetically confined nuclear outflow.

At infrared wavelengths, Doppler-broadened spectral lines of neutral helium and singly ionized neon are observed close to the nucleus. The emitting gas has corresponding velocity dispersions of 400 kilometers per second within the central 10 light-years, and of 1,400 kilometers per second within a diameter of only 1 light-year. If gravitationally bound, these motions indicate a gravitating mass of 1–10 million Suns inside a cubic light-year of volume—or about 10^{11} times the mass density of the Solar Neighborhood. Astronomers have recently traced the orbits of individual sources near the nucleus, and arrived at an estimate of 2.6 million Suns for the putative black hole that lurks within. Although such a pile-up does not require the presence of a super-massive black hole, it is difficult to conceive of any stellar or gaseous structure that could withstand the concomitant tides and radiation fields.

Imaging of the infrared continuum emission has revealed an extensive stellar cluster, with a surface brightness that falls off as the inverse square of the observed radius. The total mass of stars within a radius of 150 light-years is estimated at 10–100 million Suns—far exceeding the mass of any single globular cluster in the Galaxy. Researchers have found a similar mass for this region from the kinematics of the 21-centimeter line-emitting hydrogen gas, thus corroborating the heavyweight nature of the stellar nucleus. The relationship between this monstrous star pile, the gaseous whirlpool, and Sgr A* has yet to be worked out in any definitive way.

Our visible view of the nucleus is hopelessly obscured, but at much shorter wavelengths and correspondingly higher photon energies, the nucleus once again becomes detectable. For example, observations of X-ray emissions at wavelengths of less than 10 Angstroms indicate nuclear temperatures of 10^8 degrees Kelvin. If the emission is from a thermalized plasma, the observed low luminosity would require a source or sources no larger than 10 light-years in diameter. At even shorter X-ray wavelengths and correspondingly higher photon energies, temperatures of 10^{10} degrees Kelvin are required. The observed flux of these X-ray photons further reduces the size of the emitting source(s) by another factor of 100. The overall X-ray luminosity of the nucleus is surprisingly low. If a super-massive black hole was accreting its surroundings, there would be copious amounts of X-ray emission. Perhaps we are seeing such a black hole, but during one of its quiescent phases. Recent observations with the Chandra X-ray Observatory have confirmed short timescale fluctuations in the nuclear X-ray emission—consistent with the picture of a compact, low-luminosity accretion disk surrounding the central black hole.

Unlike the X-ray observations, the higher-energy gamma-ray observations appear to indicate the underlying presence of a very small yet intrinsically powerful nucleus. Gamma-ray emission from electron-positron annihilations has been detected. The gamma-ray source has a luminosity of 10,000 Suns and varies strongly on timescales less than a third of a year. To avoid smearing out this variability, light-travel considerations require the source to be no larger than a third of a light-year across. Gamma-ray emission at even higher energies from the decay of the isotope ^{26}Al to ^{26}Mg (the superscript referring to the total number of neutrons and protons in the nucleus) has also been detected. Here, the decay could be the result of thousands of supernovae all exploding within a million years of one another, or of just a few powerful explosions from a super-massive object. A black hole of a million solar masses would once again fill the bill.

If our imaginary giant moth could penetrate the bulge and enter the realm of the nucleus, what would it experience? Would it be able to follow the brilliant light to its source, or would it be blown away by magnetized winds long before it could get there? Would it find a blinding hot accretion disk at the core, or would its view be dominated by a stellar pile-up of Olympian proportions? And would it see the nucleus as a strongly interacting member of the Milky Way, or as an isolated vestige of what was once a feeding quasar? With these musings, we had best leave our imaginary mega-moth to its Galactic fate and begin to seek new insights from beyond the Milky Way.

7

THE CLOUDS
OF MAGELLAN

Invisible from northern latitudes above 20 degrees, the Clouds of Magellan have been familiar to southern navigators for centuries (see Figure 7.1). Fifteenth-century sailors knew them as the "Cape Clouds" and found them a useful navigational aid. South of the equator, where the north star, Polaris, is perpetually below the horizon and no bright star to mark the South Pole can be found, the Magellanic Clouds make a nearly equilateral triangle with the South Pole and thus serve as a crude guide. Their use in this way was already common when the great circumnavigator Ferdinand Magellan made his epochal voyage in 1518–1520. Although Magellan died prematurely in the Philippines, his ship continued on to Europe, where his associate and the journey's official recorder, Antonio Pigafetta, suggested that the Cape Clouds be called the Clouds of Magellan as a sort of memorial.

On a dark clear night away from city lights, a viewer sees the Large Magellanic Cloud subtending about 5 degrees, about 10 times the apparent diameter of the Moon. In similar conditions, the Small Magellanic Cloud appears to span about 2 degrees. These measurements include only the brightest portions of the Clouds, of course. On sensitive wide-field images, one can trace the Large Magellanic Cloud out to a diameter of over 10 degrees, and the Small Cloud to 6 degrees or more.

The two Clouds are sufficiently bright that the casual skygazer can mistake them for dimly lit terrestrial clouds. Photometric measurements indi-

FIGURE 7.1   The Large and Small Magellanic Clouds share the sky with the southern Milky Way in this view from Cerro Tololo Inter-American Observatory, located in the Andean foothills of Chile.

cate that if the diffuse light from the Large Cloud (LMC) was concentrated into a point, the resulting source would appear to be among the dozen brightest stars in the sky. The Small Cloud (SMC) is about four times fainter. Both are quite blue in color, owing to the considerable number of very bright, blue stars that they contain (see Plates 14 and 15).

The Magellanic Clouds are fairly typical of their class—irregular galaxies with a minimum of structure. Each has a brightest portion and various irregularly distributed small segments of similarly bright areas. Neither galaxy shows evidence for a central bulge or an active nucleus. Instead, the LMC is dominated by a bright, long, linear structure that resembles the bars seen in barred spiral galaxies. The SMC has a porkchop-shaped central core instead of a bar. Asymmetry is the rule.

Despite their irregularity, the Magellanic Clouds are not chaotic. Both show motions that are relatively well organized. The LMC, particularly, shows a regular rotational motion that resembles that of spiral galaxies such as the Milky Way. The speed of rotation is slow, reflecting the small overall mass of the galaxy. Masses are estimated to be about 20 billion Suns for the LMC and 5 billion Suns for the SMC. By comparison, our Galaxy is estimated to have a mass of over 400 billion Suns. (These figures refer to the main bodies of these galaxies; all may be grossly underestimating the total mass, because of large amounts of unseen matter that may exist in the outer parts; see Chapter 5).

Bountiful Clusters

The LMC alone is thought to have about 6,500 star clusters, and the SMC nearly 2,000. This wealth of clusters includes a wonderful variety of objects, from giant assemblages of a million stars to faint, tiny aggregates of a dozen faint dwarf stars (see Figure 7.2). The Magellanic Clouds, in fact, show an even more profuse range of clusters than does our own Milky Way, a fact that led to early confusion about what kind of cluster was what.

In our Galaxy, the globular clusters are usually large, ancient, and deficient in heavy elements, while the open clusters are almost always small, relatively young, and like the Sun in chemical abundances. The Galactic globulars typically span hundreds of light-years, contain 10^5–10^6 stars, and look like huge, bright swarms of stars, while the open clusters are only 2 to 10 light-years across, contain at most a few hundred stars, and are often too inconspicuous to discover easily.

In 1930, with the Milky Way's clusters as a reference, Harlow Shapley initially divided the clusters in the Magellanic Clouds into these two types; some were clearly globulars but the majority seemed to be tiny, poor open clusters. But even in the 1920s there was a confusing observation. Annie Jump Cannon at Harvard had examined the spectra of several of the globular clusters on Shapley's lists and found some of them to be much bluer than globular clusters in the Milky Way. This fact was puzzling at the time because stellar evolution was not yet understood; we now know that these clusters' spectra indicated the presence of hot, young stars.

Decades later, photoelectric measurements verified the true colors of the Magellanic globulars' colors. The brightest clusters fell into two groups: "normal" red globulars and "abnormal" blue globulars. After many more years of research, astronomers now recognize that the blue globulars are young clusters that are otherwise large and populous like the Milky Way's old globular clusters. They range in age from fledgling objects like the giant

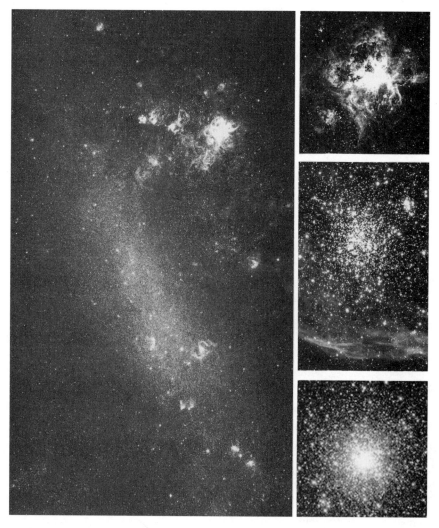

FIGURE 7.2 The Large Magellanic Cloud (left) contains many star clusters and associated nebulae. Top right: The Tarantula Nebula (also known as 30 Doradus), one of the largest and most powerful starburst regions in the local Universe; powered by the brightest and most short-lived O-type stars in the underlying cluster, the Tarantula is just a few million years old. Middle right: The cluster NGC 1850; it contains stars of type B and cooler (the hotter and more massive O-type stars having died off), which indicates an age of about 50 million years. Bottom right: The cluster NGC 2019; it contains only the longest-lived stars—main-sequence and giant stars of type K and M—indicating a hoary age of 15 billion years.

NGC 1850, a mere 50 million years old, to clusters of advanced middle age like NGC 1978, which contain stars with lifetimes of a few hundred million to a billion or so years. For the older ones, especially, the difficulties in age-dating the clusters were not overcome until telescopes with apertures of 3.6 to 4 meters were built in the Southern Hemisphere during the 1970s. Only then was it possible to reach down to the faint, unevolved stars that must be measured to reliably gauge a cluster's age.

Only a handful of genuinely old globular clusters have been found in the Clouds; the SMC has one, NGC 121, and the LMC has about 12, of which NGC 2019 is a good example (see Figure 7.2). Most of them contain variables of the RR Lyrae type, which in the Milky Way are associated with the ancient, metal-poor population of globular clusters that inhabit the Galactic halo. Like the Milky Way globulars, these clusters have ages measured to be about 12–14 billion years—about as old as the Universe. The rest of the 40 or so large, red, populous clusters apparently are younger.

Unlike our Galaxy, the Magellanic Clouds have been able to produce giant, rich clusters right up to the present. For reasons not yet thoroughly understood, the Milky Way stopped making such clusters around 13 billion years ago, and since then has just been able to construct small clusters, which in a billion years or less are torn apart by tidal forces and other disruptive effects.

The clusters of the Clouds illustrate the dramatic differences between these two irregular galaxies and our own system. The Magellanic Clouds are loose, low-mass, slowly rotating systems that are not strictly confined to a flat plane and that probably do not have a large spherical halo of old stars, as does the Milky Way. In this more gentle environment, star clusters can form and prosper in ways that are impossible in the tidally strained disk of our much more massive Galaxy.

The wide range in age of rich Magellanic clusters is of special interest to stellar theorists. Because the open clusters in the Milky Way are rather poorly populated—having only a few, if any, stars that happen to be passing through the later, giant phases of evolution—we are hampered locally when we want to compare our theories of stellar development with real stars. But the rich Magellanic Cloud clusters provide lots of stars that are going through their various giant phases—expanding, contracting, and expanding again, before finally collapsing to obscurity as white dwarfs or neutron stars.

Like the best of teachers, the Cloud clusters have revealed to astronomers what happens in real life, providing tangible benchmarks for their sundry theoretical calculations. Given that the basic relations between a star's mass, composition, age, size, luminosity, and temperature depend on the particular stellar model, it is of great importance to have a check on the theory. Astronomers want to be sure that the various necessary assumptions and approximations that go into such a mathematically complex theory have not led them astray. So far, the comparisons between stellar evolutionary theory and the stars observed in the Magellanic Cloud clusters have yielded remarkably favorable correspondences.

For example, the clusters of the Clouds have provided important clues to understanding the relations between stellar age, chemical composition, and the masses of the galaxian hosts. In our own Galaxy, as we have seen, there is a correlation between the ages of clusters and the chemical abundances in their stars. The very oldest clusters in the Milky Way, the globulars, tend to have stars that are deficient in elements heavier than helium and hydrogen, such as calcium, iron, and magnesium. Younger stars—for example, stars in open clusters like the Pleiades—have a richer mix of these elements, about the same amount as in the Sun. This is understood as being the result of the gradual enrichment of heavy elements in our Galaxy as stars evolve. A star in its giant phases will produce carbon and, perhaps, some nitrogen in its core, while very massive stars that explode into supernovae—destroying themselves in a brief moment of glory as bright as a billion Suns—produce small amounts of all the heavier elements. The longer a galaxy exists and the more supergiants and supernovae it has hosted, the more heavy elements are dispersed into its interstellar medium. Succeeding generations, then, become richer and richer in heavy elements.

For the Clouds, the environment is different, and the question arises as to how this might have affected their chemical history. When we look at very young objects in the Clouds, like H II regions or young stars, the spectra suggest that the *current* abundance of heavy elements is less than that in the Sun. The SMC is low by about a factor of five and the LMC by about two or three. Because individual clusters can be age-dated reliably, it is possible to trace how this circumstance has come about. The conclusion is that both Clouds seem to have started out, about 15 billion years ago, with the same low heavy-element abundance that our Galaxy had back then. But in the years that followed, heavy elements built up more gradually, especially for the Small Cloud, ending in the paltry values observed today. They appear to be the result of different average rates of star formation and destruction. The lower-mass galaxy, the SMC, processed stars most slowly, while the higher-mass galaxy, the Milky Way, processed them most rapidly. By extrapolation, one might guess that even lower-mass galaxies, such as IC 1613, should have even lower present-day abundances of heavy elements, and this does seem to be true (see Chapter 8). Similarly, very massive galaxies like the brightest seen in the Virgo cluster ought to be unusually rich in heavy elements, a result that is also confirmed by the metal-rich spectra of these systems.

Evolutionary Tales

The star clusters in the Clouds have also provided a means to study the evolutionary histories of the Clouds themselves. It is not easy to trace the history of a galaxy between its formation and the present, because so much depends upon the wide variety of gravitational and other forces at work in different parts of the galaxy. We must employ methods that are a little bit like archaeology: we start at the present and dig down, layer by layer, to see if a pattern emerges that will give us clues regarding the life history of the galaxy.

The goal is to find the rate of star formation at different times and at different places in a galaxy. One way to do it is to look at the distribution of star clusters of different ages in the Magellanic Clouds. In the LMC, for example, cluster formation seems to have occurred sporadically and preferentially in groups, with the mean population for groups being approximately 25 clusters, the mean diameter of the groupings of clusters being approximately 2,000 light-years, and the mean duration of enhanced cluster formation being several million years. We can also see that the big, massive clusters were formed even more sporadically than the small, open clusters—with preferences for certain epochs: 12–15 billion years ago and 1–2 billion years ago, for instance.

Much of this star-forming history can be visualized via three-dimensional diagrams (see Figure 7.3). In these so-called population boxes the horizontal, vertical, and diagonal axes respectively refer to the ages, star-formation rates, and metal abundances in the galaxies. Each plotted value in the boxes has been derived from painstaking observations of the stars and star clusters in the galaxies.

The resulting surfaces in these three-dimensional population boxes indicate the rates at which the galaxies have converted their primordial gas into stars and, in so doing, have enriched their remaining gas with elements heavier than helium. A quick examination of the boxes in Figure 7.3 reveals important differences in the evolutionary histories experienced by the Milky Way galaxy, the Andromeda galaxy, and the Large Magellanic Cloud. Both of the giant spirals began early on to make lots of stars. Consequently, their chemical abundances ramped up to the high values we see today near the Sun. The bulges of these giant systems evolved faster than their outer parts, however, and this explains the broad range of abundances at any given time. By contrast, the LMC held back in its star formation, so that it is still as active today as it was some 15 billion years ago. The bursts of star formation 15

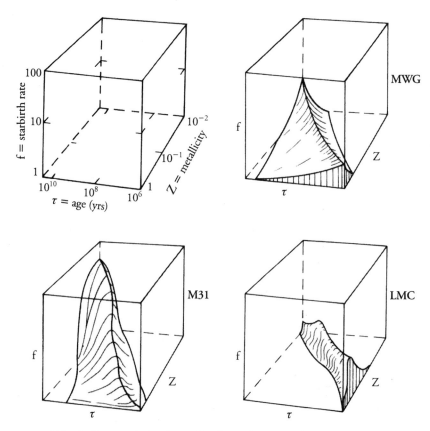

FIGURE 7.3   Stellar population boxes for the Milky Way galaxy (MWG), Andromeda galaxy (M31), and the Large Magellanic Cloud (LMC). By mapping both the star formation rate (f) and chemical abundance (Z) as a function of age (τ), astronomers can trace a galaxy's evolutionary history. The giant spirals (MWG and M31) show rapid declines in starbirth activity and robust chemical enrichment, while the LMC shows a fairly steady progression of star formation with modest chemical evolution.

and 1 billion years ago further serve to differentiate the LMC from its larger brethren. The LMC's spread of chemical abundances is considerably less than that evident in the Milky Way and M31—another testament to its lower mass and less nucleated structure.

The Brightest Stars

Much of our own Galaxy is hidden from us. Optically, we can penetrate into the thick veil of dust that pervades the Milky Way only a few thousand light-years in most directions before the absorption and reddening of intervening dust clouds obscure more distant stars. For that reason, the Magellanic Clouds are an important source of information about the rarest types of stars, objects that are too scarce for there to be many visible in our own system.

The most luminous stars in a galaxy are the rare supergiants, stars that are 10^5–10^6 times as luminous as the Sun. A few of these extreme objects have been discerned in the murky disk of the Milky Way, but we have nothing like a complete census. In the Magellanic Clouds, which are spread out clearly for us to sample, and which have far less dust for us to contend with, we can see virtually all the supergiants; we can study them and deduce their histories, their remarkable makeup, and their life expectancies. We can learn about the evolution of a galaxy's most massive stars, and we can test the possibility of using these brilliant objects to gauge the distances to far more remote galaxies, where only the very brightest stars are resolvable (see Chapter 12).

Modern study of the most luminous supergiants in the Magellanic Clouds began in the 1950s, when photoelectric data were combined to identify supergiant stars among the plethora of foreground stars in the direction of the Clouds. It is not a trivial problem. The stars are so rare that it is not easy to separate them from the stars of our own Galaxy, which are more numerous at these levels of apparent brightness, though intrinsically far less luminous.

How can an astronomer tell whether a given bright star is a supergiant member of the Clouds or merely a normal-sized star in the foreground, seen by chance in front of the Clouds? The first clues came many years ago at Harvard, when Cannon's spectroscopy in the direction of the Clouds showed the presence of very blue stars, including certain peculiar stars with envelopes of expanding gas, known as Wolf-Rayet stars. These kinds of stars are among the most luminous of the local stars in the Milky Way and are found very close to the midplane of the Milky Way, almost never at positions in the sky as far from the plane of the Milky Way as the Clouds lie. It could be argued, therefore, that the stars Cannon had detected must belong to the Clouds and be among their brightest members.

This spectroscopic method of determining membership in the Clouds has since been extended to include medium- and high-dispersion spectroscopic studies. At intermediate resolutions, the mean velocities of hundreds of prospective stars were measured to weed out those whose small radial velocities indicated that they are nearby stars instead of members of the Clouds (the radial velocities of the Clouds with respect to the Sun are large, 168 kilometer per second for the SMC and 276 kilometers per second for the LMC). At high resolution, one can directly determine whether or not a given star is a supergiant from the Doppler-broadened width of its spectral lines, such that the distended supergiant atmospheres yield lower pressures and hence narrower linewidths. The resulting census of supergiant stars has shown that there are thousands of supergiant members in the Clouds.

The visually brightest stars in the Clouds are actually not the hottest, as one might expect. Instead, they are the A-type stars, with temperatures of about 10,000 K. The very hottest stars are 10 times fainter in visual light, emitting most of their radiation at shorter ultraviolet wavelengths; these are young, blue stars of types O and B (see Figure 7.4). The coolest supergiants, those with bloated outer envelopes of thin gas, are about half as bright visually as A-type stars. All of these supergiant types represent the final gasps of the most massive stars in a galaxy, those 20 to 100 times as massive as the Sun. Such rarities have spent most of their short lives as hot, main-sequence O-type stars, before they expanded into their fleeting supergiant phases. Thereafter, they quickly evolve through their hot and cool phases at nearly constant total luminosity, until their cores implode into neutron stars or black holes, and their surrounding layers explode as brilliant supernovae.

Because most of the radiation emitted by hot stars appears in the ultraviolet part of the spectrum, most of the light from these stellar powerhouses is absorbed by the Earth's atmosphere. Thus to record most of the radiation from these very luminous, high-temperature stars, astronomers must use orbiting instruments that view the stars from above the atmosphere. A close ultraviolet examination of the Magellanic Clouds' brightest stars has helped to separate the hot, main-sequence stars from the evolved supergiants of similar temperature.

The peculiar absorption-plus-emission–line profiles found in the spectra of many hot supergiants indicate ongoing eruptions, with mass-loss rates as great as 1 Sun every 100,000 years. These rates, though profligate, are somewhat less than those of their counterparts in the Milky Way. The less extreme loss of material from the Cloud stars is a probable consequence of the lower chemical abundances that are extant. With fewer heavy elements and dust grains in the stellar atmospheres, the emerging UV radiation has fewer "targets" to hit and so drive a superwind. Further study of the most massive stars in the Clouds has begun to provide a cursory sketch of their evolutionary sequencing from type to type—and of their impressive powering of the surrounding interstellar matter.

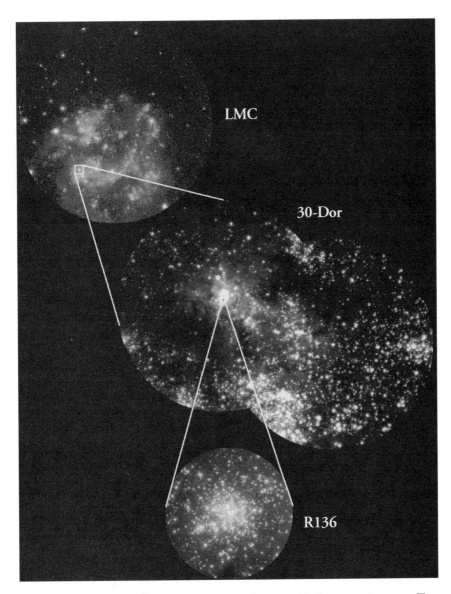

FIGURE 7.4   Ultraviolet views of the LMC are dominated by hot supergiant stars. Top: Spanning 30,000 light-years, the LMC contains a multitude of UV bright star clusters. Middle: The 30-Doradus region, 3,000 light-years across, includes several associated clusters of hot young stars. Bottom: The dense cluster R136 within 30-Doradus is only 30 light-years wide, yet contains thousands of massive hot stars.

Cepheid Variables

Historically, the Magellanic Clouds' most important role has been as a realm for the discovery and testing of the relation between luminosity and cyclic period in certain stars whose brightness varies with time. More than 10,000 variable stars have been found in each Cloud, most of them of the type called Cepheids. These bright yellow supergiant stars show regular variations in brightness with periods of oscillation on the order of days. Following Henrietta Leavitt's discovery of the period-luminosity relation in Cepheids early in the twentieth century, other astronomers, especially Harlow Shapley, became involved in the study of Cepheid variables. He emphasized their importance in distance measurements, and derived the Cepheid's period-luminosity relation for the Clouds.

As more reliable photometric data on these stars accumulated, the exact shape of the period-luminosity relation was refined (see Figure 7.5). It is on this kind of baseline data that a distance scale of the Universe would eventually be built (see Chapter 12). The Magellanic Clouds, being the nearest galaxies, were and continue to be the first vital step in this effort.

Great Clouds of Glowing Gas

On just about any photograph of the Large Magellanic Cloud, one of the most conspicuous features is a bright, sprawling object, located just east of the center of the main bar of stars (see Plate 14 and Figure 7.2). This is one of the most massive and luminous hot clouds of gas (H II regions) that we know of in any normal galaxy. It is the Tarantula Nebula, also called 30 Doradus, the name of the constellation in which it is located. Illuminated by a dense cluster of hot stars within it, the Tarantula Nebula glows with the light of millions of Suns.

Although similar in nature to the Orion Nebula in the Milky Way, the LMC nebula is immensely larger and more luminous. While the Orion Nebula is about 40 light-years across, 30 Doradus extends across a thousand light-years and is a thousand times brighter. If it were situated in our Galaxy at the position and distance of the Orion Nebula, it would spread out over the entire constellation and would be bright enough to cast shadows on the Earth. We have no such immense object in the Solar Neighborhood; only by radio surveys of the Milky Way have we found any H II regions approaching the size of the Tarantula Nebula, and then only at great distances, hidden deep within the Galaxy's disk.

The total mass of the gas in 30 Doradus is about 5 million times that of

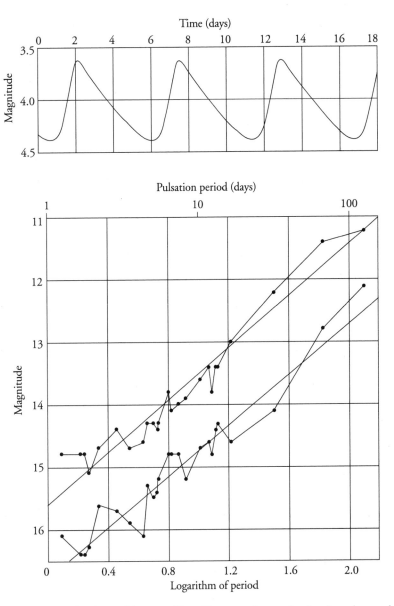

FIGURE 7.5 Cepheid variable stars. Top: Photometric monitoring has shown that Cepheid variables brighten and fade over time in a characteristic sawtooth pattern with pulsation periods of several days. The light curve of the prototype Cepheid variable star, Delta Cephei, is shown here. Bottom: Henrietta Leavitt's original plotting of the relationship between the pulsation period and luminosity among Cepheid variable stars (filled black circles) in the Small Magellanic Cloud. Here, the period ranges from 2 days to 100 days, and the brightness (expressed in magnitudes) ranges over a factor of 100 in actual luminosity. Each star is represented twice, at its maximum brightness (upper line) and at its minimum brightness (lower line).

the Sun. This includes the visible, glowing gas and a surrounding accumulation of neutral gas, detectable only at radio wavelengths. The optical structure of the nebula is marvelously intricate, with interwoven loops and rings extending out from its bright core. Some of these are probably the result of supernova explosions and others are probably formed by the intense winds of gas emitted by high-luminosity stars.

Embedded within 30 Doradus is a rich cluster of hot supergiants known as R136 (see Figure 7.4). The central core of R136 contains a remarkable object of tremendous luminosity, called R136a. Using the Hubble Space Telescope, as well as groundbased telescopes equipped with imagers that compensate for the atmospheric distortions, astronomers have resolved this kernel into a dozen or so stars that are very close to each other. These are some of the most massive stars ever found anywhere and, packed together as tightly as they are, they form a blazing core for the Tarantula—brighter and more massive than most theoretical models would predict. Once again, nature seems to be saying that our imaginations are too limited; the schemes we devise are seldom so grand as what nature has already accomplished.

There are thousands of other glowing gas clouds in the LMC and hundreds are catalogued for the SMC. The diverse structures of these H II regions help astronomers to better understand the dynamics of expanding gas clouds in different environments. The H II regions in the Clouds also provide important information on the evolving chemical composition of the interstellar medium. Compared with the gas in the Milky Way (the Orion Nebula, for example), the gas regions in the Clouds are less abundant in heavy elements. Even helium is noticeably deficient, the result of there having been fewer generations of stars in the Magellanic Clouds. While in the Sun's neighborhood a given atom (such as one inside this page or in your body) might have, on the average, existed at various times in three different stars, in the Magellanic Clouds a typical atom will have had a less interesting past. Perhaps it will have had only one previous stellar incarnation.

Dust in the Clouds

Because dust is made largely of heavy elements, it might be supposed that this material should be rarer in the Magellanic Clouds than in our dusty Galaxy; and that does indeed seem to be the case. Several discrete "dark nebulae," as they are often called, have been catalogued in each Cloud, and they seem to have normal properties when compared with dust clouds in the Milky Way. But the total number of discrete dust clouds in the Clouds is much smaller than in our Galaxy.

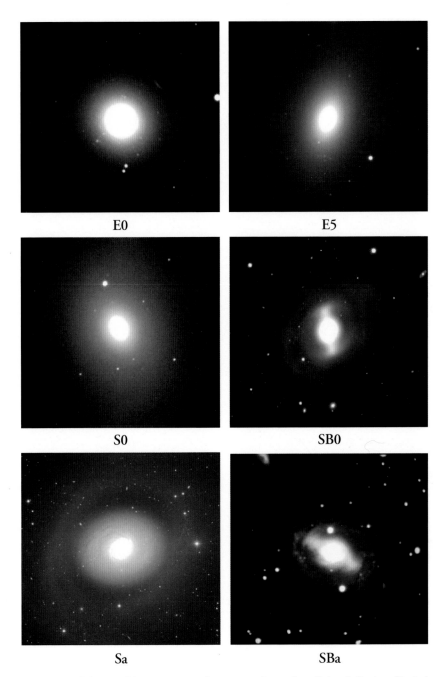

E0   E5

S0   SB0

Sa   SBa

PLATE 1   Galaxies with strong central concentrations of starlight. Left: An elliptical (E0), a lenticular (S0), and a spiral of type Sa. Right: Elongated or barred (B) versions of their respective types. From top to bottom, the galaxies are NGC 4552 (M89) and NGC 4621 (M59), NGC 4382 (M85) and NGC 936, and NGC 4736 (M94) and NGC 4650.

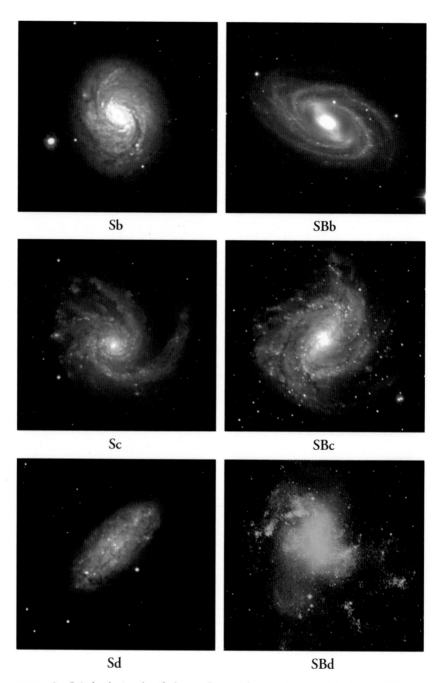

Sb

SBb

Sc

SBc

Sd

SBd

PLATE 2   Spiral galaxies classified according to the prominence of their central bulges and the tightness of their spiral arm windings, the Sd spirals having the weakest bulges and the loosest spiral structure. From top to bottom, the galaxies are NGC 1068 (M77) and NGC 3992 (M109), NGC 4254 (M99) and NGC 5236 (M83), and NGC 2976 and NGC 1313.

Sm

SBm

Im

Amorphous

Interacting

Peculiar

PLATE 3   Irregular galaxies of Sm, SBm, and Im types, along with amorphous, interacting, and peculiar galaxies that defy easy classification. From top to bottom, the galaxies are NGC 4449 and the Large Magellanic Cloud, the Small Magellanic Cloud and NGC 3034 (M82), and NGC 4676 (the Mice) and NGC 4650A.

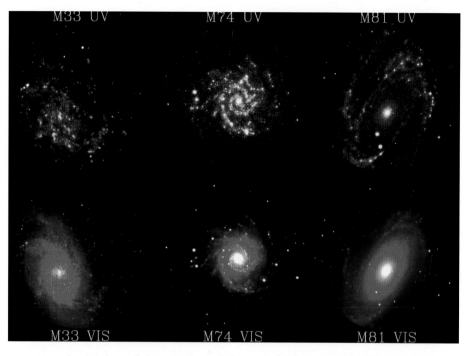

PLATE 4 A color-coded comparison of three spiral galaxies as imaged in ultraviolet (top) and visible (bottom) light. Hot young stars in the spiral arms dominate the ultraviolet view, while cooler and typically older stars in the disks and bulges prevail at visible wavelengths.

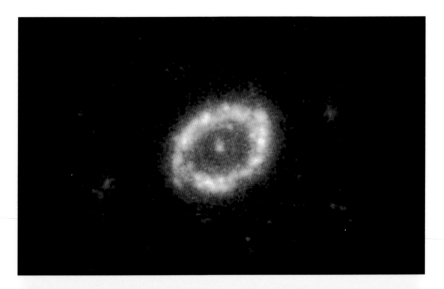

PLATE 5   Resonant-ring Sab galaxy M94 (NGC 4736) as imaged in ultraviolet (top) and visible (bottom) light. The colorized ultraviolet image shows clear evidence for a ring of hot young stars created as part of recent starburst activity there. The visible view highlights the 10-billion-year-old legacy of cooler stars that have accumulated in the disk and bulge.

Visible

Near-Infrared

PLATE 6 All-sky mapping in visible (top) and near-infrared (bottom) light. The Milky Way extends across the middle, with the Galactic center situated in the exact center of each mapping. The Large and Small Magellanic Clouds, companion galaxies to the Milky Way, appear to the lower right.

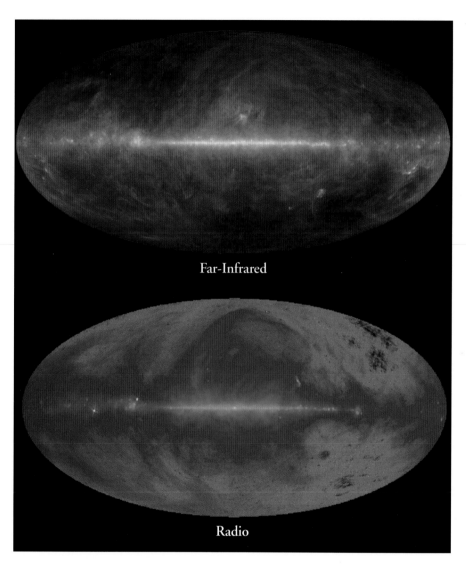

Far-Infrared

Radio

PLATE 7   All-sky mapping in far-infrared (top) and radio (bottom) light. The colorized far-infrared mapping reveals myriad clouds of star-warmed dust in the Milky Way. The colorized map of radio emission is dominated by ionized gas in the Milky Way, including a large supernova-driven bubble fragment above the Galactic plane.

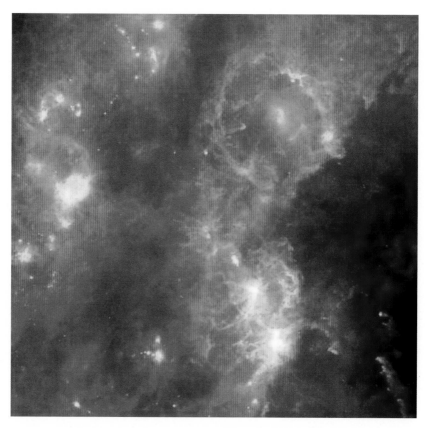

PLATE 8   Far-infrared emission from stellar nurseries in the direction of Orion the Hunter. The field of view is roughly 30° × 40°. The Orion Nebula region appears at the bottom, with the Horsehead Nebula region just above it.

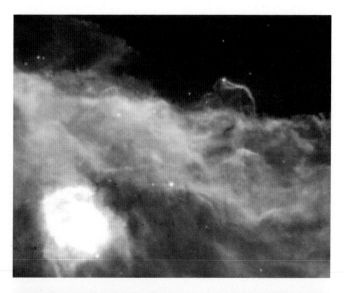

PLATE 9 The Horsehead Nebula at mid-infrared (top) and visible (bottom) wavelengths. The colorized mid-infrared view showcases emission from irradiated dust grains and organic molecules in the cloud. The true-color visible view highlights the obscuring effects of dust associated with the protruding molecular cloud.

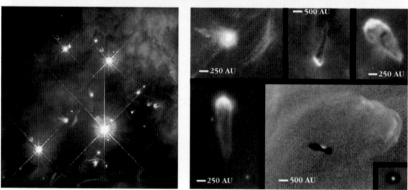

PLATE 10 The Orion Nebula with closeups of its stellar core. Top: The fluorescing nebulosity (H II region), a scene spanning 2.5 light-years. Bottom left: The Trapezium core of hot O-type stars, dominated by Theta Orionis C. Bottom right: Cometary proto-stars and protoplanetary disks (proplyds) near Theta Orionis C, rendered with exaggerated colors. The size scales are in Astronomical Units (AU), where 1 AU is the distance between the Sun and the Earth.

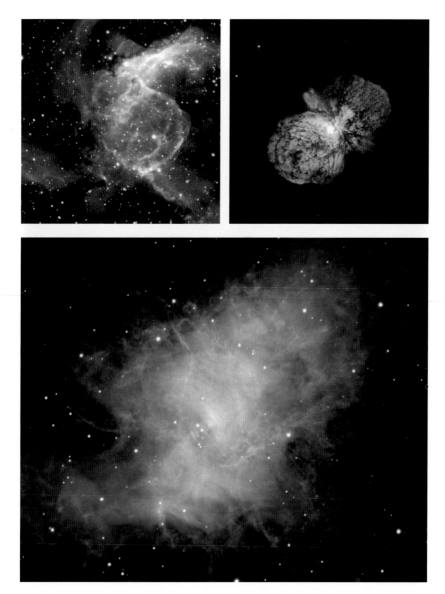

PLATE 11 Stellar eruptions. Top left: The nebulous bubble NGC 2359 was blown out by the massive and extremely hot Wolf-Rayet star near its center. Top right: The massive and luminous blue variable star Eta Carinae suffered a major eruption in the early 1800s. Bottom: The "Crab" supernova remnant (M1) is the nebular consequence of a stellar explosion that was recorded on Earth in 1054 C.E.

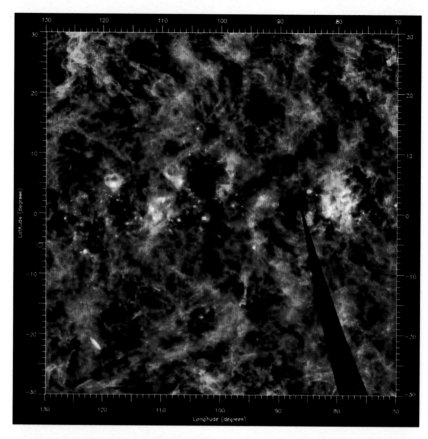

PLATE 12 A 60° × 60° swath of the Milky Way near the Cygnus star-forming region, as viewed at far-infrared wavelengths. This processed and colorized image highlights the frothy structure of the interstellar medium. Star-forming regions punctuate the Galactic midplane. The Andromeda galaxy (M31) is evident to the lower left. The wedge of darkness indicates a zone that was not mapped by the Infrared Astronomy Satellite during its survey of the sky.

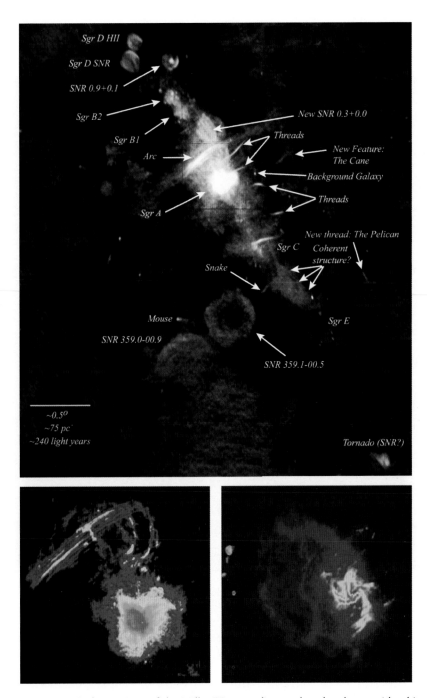

PLATE 13 Nuclear regions of the Milky Way at radio wavelengths, shown with arbitrary colors. Top: The central 1,000 light-years of the Milky Way contain many supernova remnants (SNRs) and H II regions along with strange "threads" of gas. Bottom left: The central 250 light-years include the dense Sagittarius A core region and arcs of magnetized gas. Bottom right: The innermost 25 light-years feature spiral streamers of ionized gas that converge on the nucleus, Sagittarius A*, where a super-massive black hole is thought to dwell.

PLATE 14   The Large Magellanic Cloud (LMC). The bright red object near the top is the Tarantula Nebula (also known as 30 Doradus), a supergiant H II region the equivalent of 1,000 Orion Nebulae.

PLATE 15   The Small Magellanic Cloud (SMC).

PLATE 16   The inner disk and bulge of the Andromeda galaxy, M31 (NGC 224), along with the smaller satellite galaxies M32 (right) and NGC 205 (left).

PLATE 17   The Pinwheel galaxy, M33 (NGC 598). The many crimson nebulae in M33 are giant H II regions energized by hundreds of newborn hot stars inside them.

PLATE 18 Color-coded views of the grand-design spiral galaxies M81 (top) and M74 (bottom). The blue-white colors along the spiral arms indicate regions of bright ultraviolet emission from hot young stars. The yellow-red colors show the underlying disks of cooler and typically older stars.

PLATE 19   The grand-design barred spiral galaxy M83 (NGC 5236).

PLATE 20 Circumnuclear starburst rings in NGC 4314 (top) and NGC 7742 (bottom).

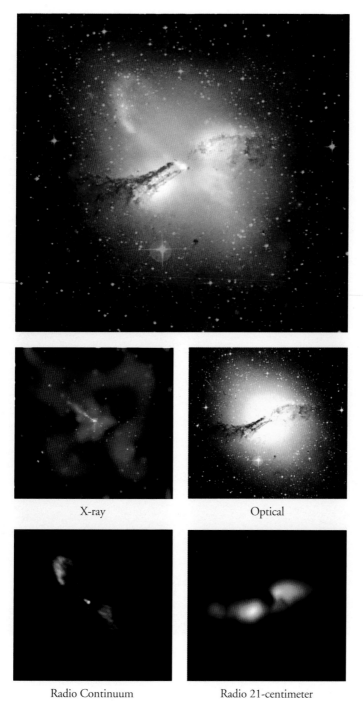

X-ray   Optical

Radio Continuum   Radio 21-centimeter

PLATE 21   Centaurus A (NGC 5128), a giant elliptical galaxy in the process of devouring a spiral galaxy and blowing out vast jets of gas from its nucleus. The color-coded composite image (top) shows the elliptical galaxy girdled by dusty remnants of the spiral galaxy and enveloped by hot gas. The smaller frames (bottom) show the gaseous and stellar components at X-ray, optical, radio-continuum, and radio 21-centimeter wavelengths.

PLATE 22 The Whirlpool galaxy, M51 (NGC 5194 and its interacting companion NGC 5195). The magnificent spiral structure in M51 was probably induced by one or more close passages of the smaller amorphous galaxy.

PLATE 23   The Antennae involve two spiral galaxies (NGC 4038 and NGC 4039) that have been severely distorted from their merging encounter. The closeup view of the region outlined in green at the top reveals enormous clusters of newborn hot stars, each the equivalent of several thousand Orion Nebula clusters.

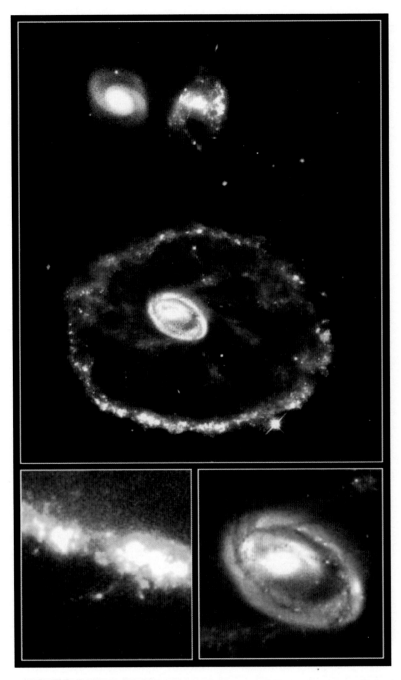

PLATE 24 The Cartwheel (middle) was probably created from a head-on collision between a small galaxy and a much larger spiral galaxy. Closeups (bottom) highlight the starburst activity in the ring (left) and the disturbed nature of the Cartwheel's remnant core (right). The small galaxies (top) are not thought to have been directly involved with this violent event.

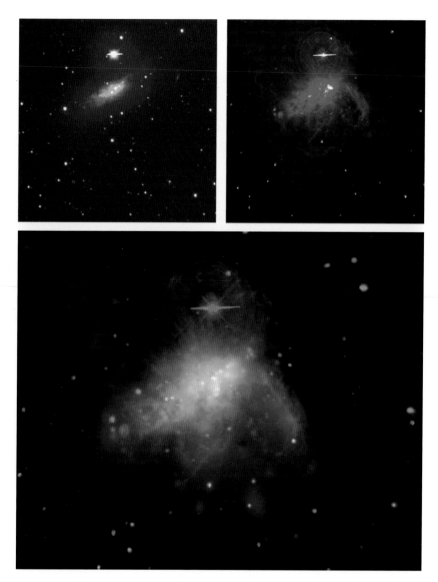

PLATE 25   The nearest starburst galaxy, NGC 1569, appears in visible light as a modest blue irregular system (top left). In the red light of excited hydrogen gas (top right), a big blowout is evident. The composite, color-coded scene (bottom) shows million-degree X-ray–emitting gas (in green) filling up the volume enclosed by the relatively cooler and denser hydrogen gas.

PLATE 26 The amorphous galaxy M82 (NGC 3034), with its starbursting nucleus. Red filaments of excited hydrogen gas outline the superwinds that are emanating from the nuclear starburst. These winds were generated by the explosion of several hundred thousand massive stars in the core of this galaxy within the last few million years.

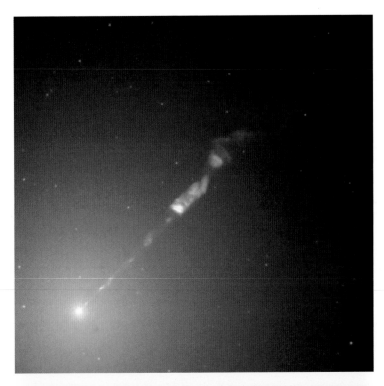

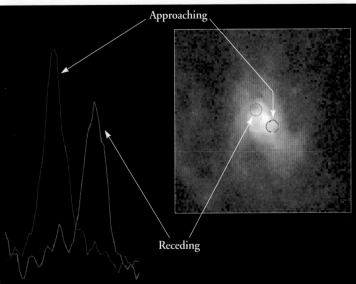

Approaching

Receding

PLATE 27 The giant elliptical galaxy M87, showing its inner jet (top) and nuclear accretion disk (bottom). Blue and red shifts in the light from the inner disk respectively indicate rapid motions toward and away from us. These motions, if gravitationally bound, imply the presence of a super-massive black hole in the galaxy's nucleus.

PLATE 28   The Coma cluster of galaxies (top) is dominated by elliptical galaxies that contain mostly old, yellow-orange stars. A closeup (bottom), centered on the giant elliptical galaxy NGC 4881, shows one of the rare spiral galaxies in Coma plus myriad other background galaxies much farther away than the cluster itself. The blank region at the upper left was not imaged by the Hubble Space Telescope's camera

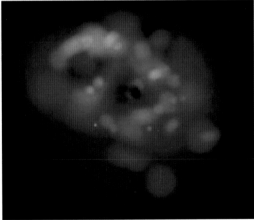

PLATE 29   The galaxy cluster CL0024+1654 has gravitationally lensed and magnified a more distant blue galaxy into many ghost images (top). A reconstruction of the undistorted background galaxy reveals a strange pretzel-like structure in the galaxy's ultraviolet-emitting population of young hot stars (bottom).

PLATE 30. The Hubble Deep Field reveals thousands of galaxies in a piece of the sky measuring only a few hundredths of a degree on a side—the equivalent angular expanse that would be subtended by the eye of Franklin Roosevelt on a dime held at arm's length. If extrapolated to the entire celestial sphere, these numbers indicate the presence of about 50–100 billion galaxies in the visible Universe. The blank patch at the upper right was not imaged by the Hubble Space Telescope's camera.

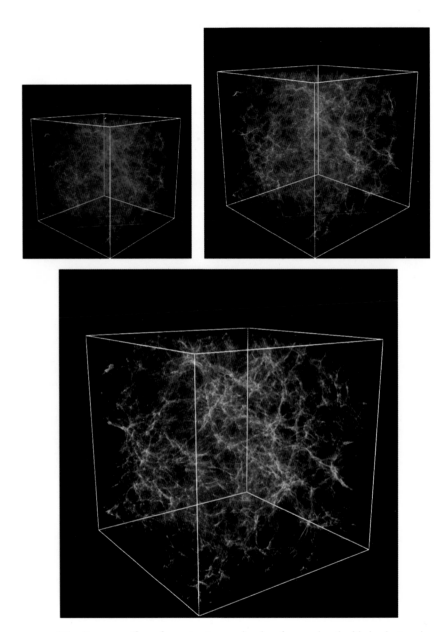

PLATE 31   From a uniform fog to intricate cobwebs of gas and embedded galaxies, the growth of large-scale structure in the Universe can be tracked by computer simulations. The displayed sequence of three frames shows a billion-light-year piece of the Universe when it was half its current size (top left), two-thirds its current size (top right), and as it is today (bottom).

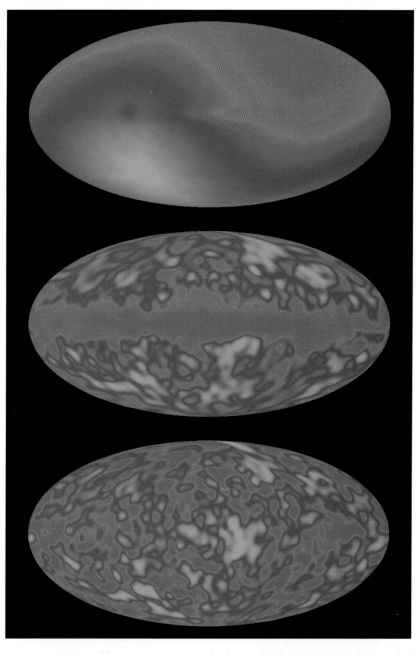

PLATE 32 Maps of the cosmic microwave background radiation as originally recorded by the Cosmic Background Explorer (top), after subtracting the bipolar variation in temperature due to our Galaxy's peculiar motion (middle), and after subtracting luminous contributions from the Milky Way (bottom). The remaining emission is thought to originate from the primeval Universe, some 300,000 years after the Big Bang.

In the SMC, the counts of background galaxies indicate that the total extent and amount of obscuring interstellar dust is rather modest. To compare various galaxies' dust content, astronomers often refer to their dust-to-gas ratio, as a way of normalizing the data. From the galaxy counts, the SMC's dust-to-gas ratio turns out to be about 30 times lower than that of the Milky Way. Compared with the SMC's five-fold underabundance of heavy elements, its allotment of dust is another 6 times less. The Small Cloud's lack of dust is also evident at far-infrared wavelengths, where cold dust naturally emits, but is barely detected. Additional processes seem to be conspiring with the overall paucity of heavy elements to inhibit the formation of dust in this small galaxy.

Further confirmation of the SMC's dust-poor state has recently come from a series of measurements of carbon monoxide (CO) obtained with radio telescopes. In our Galaxy, CO is found in giant molecular complexes, whose dust content is known to scale with the total CO emission. By comparison, the CO in the SMC is very sparse. The most sensitive millimeter-wave radio telescopes find only weak CO signals from a few of the optically catalogued SMC dust clouds. It appears that whatever is inhibiting the formation of dust is also preventing molecular gas from gaining a significant toehold in the SMC—despite the galaxy's high content of atomic gas (see below).

Optical and infrared studies of the Large Cloud suggest more dust, but still not anywhere as much as in our Galaxy. Carbon monoxide clouds are also found there, especially in areas where current star formation is prominent and dust is present. The brightest stars and Cepheids of the LMC seem to be reddened only slightly, mostly by dust in the foreground of our own Galaxy. A few are dimmed by as much as 50 percent or so, mostly in areas near the center of the Cloud where dust clouds are also conspicuous.

Common to both galaxies are their relatively low masses and metal abundances, yet relatively high proportions of young hot stars. Perhaps the intense ultraviolet radiation from these young stellar populations hinders the delicate processes that build interstellar molecules and dust grains, thereby amplifying the overall effect of lower metal abundances.

The Common Envelope

The hot, glowing clouds of hydrogen are optically conspicuous features of the Magellanic Clouds. But if we could see with long-wavelength eyes, tuned to the 21-centimeter wavelength that neutral, cool hydrogen gas emits, we would behold an even more impressive view (see Figure 7.6). Radio telescopic surveys have revealed that both galaxies contain H I gas complexes of immense size and complicated structure, in spite of the rather small total masses of the Clouds. As in other irregular galaxies, an unusually large proportion of their mass is in the form of neutral hydrogen, probably more than 10 percent in both cases. This seems reasonable, considering that the Magellanic Clouds have been less efficient than our Galaxy in forming stars, and so more of the raw material for star formation is still around.

In the 1950s, pioneering radio astronomers in Australia found that at a 21-centimeter wavelength the Magellanic Clouds are really only one object. An all-encompassing envelope of hydrogen includes both of them in its immense extent. The LMC is a complex, highly structured component on the east side of this envelope, while the SMC makes up a peculiarly shaped concentration on its west side. The bridge between is thin gas, with few detectable stars (see Figure 7.6).

The Australian astronomers also made an even more surprising discovery. Flung off across the sky is a huge, thin filament of gas, originating at the Clouds and crossing almost to their antipodes (see Figure 6.6). Called the Magellanic Stream, this ribbon of gas seems to connect several other very low-mass galaxies. The most reasonable explanation for the Stream is that it is a tidal tail or bridge that was drawn out from the Clouds during a close encounter with the Milky Way. Computer simulations of such an encounter suggest that the Clouds passed into the outer parts of our Galaxy in a near cataclysmic event about 2 billion years ago. The present distances and velocities of the Clouds, as well as the properties of the Magellanic Stream, fit this model well. Whether encounters occurred even longer ago remains a puzzle, as does the question of what is next in store for the Magellanic Clouds, billions of years hence when they may once again come too close.

X-Rays and Black Holes

A final word about the Magellanic Clouds involves their use as important testing grounds for the most extreme types of celestial objects—black holes. X-ray sources exist throughout our Galaxy, but most are faint and their optical counterparts are not always easy to interpret or their distances easy to measure. Many bright X-ray sources have been found in the Magellanic Clouds and these, of course, all have known distances, the distances to the Clouds. Thus we can determine their physical properties reliably. This is especially important for those that are in binary (double) star systems, because if we find that the dynamical mass of the X-ray–star is large, on the order of 10 solar masses or more, then it is probably a black hole, while if it is only 2

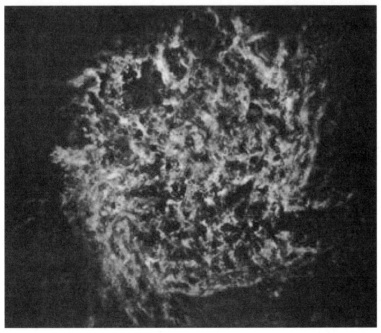

FIGURE 7.6   Neutral hydrogen (H I) in the Clouds. Top: The LMC (left) and SMC (right) are infused with atomic hydrogen gas and connected by a common bridge of diffuse H I. Bottom: The fine-scale structure of H I gas in the LMC is reminiscent of bubbly foam.

or so times the mass of the Sun, it is probably "merely" a neutron star. And to know the masses reliably, we must know the distances.

The source SMC X-1 was shown many years ago to be a binary with a neutron star, but in 1983, LMC X-3 was determined to be far too massive to be such an "ordinary" exotic object. Instead, scientists concluded that it must contain a massive black hole. The X-rays are emitted by a stream of gas that is inexorably being pulled into the black hole to disappear forever from our Universe. LMC X-3 was one of the very first black holes to be discovered, and is one more example of how the Clouds of Magellan still serve as signposts in the night.

8

DWARFS OF THE
LOCAL GROUP

Shortly after Edwin Hubble determined in the 1920s that we live in a Universe of galaxies, exploration of the galaxian realms began in earnest. Before long, thousands of galaxies were seen to the limits of all the major telescopes. Many galaxies were found among groups or clusters, beginning with our home system. The Milky Way was recognized as the dominant galaxy with respect to its close neighbors, the Large and Small Magellanic Clouds, less than 200,000 light-years away. Another giant spiral, M31 (the Andromeda Nebula), was found by Hubble to lie about 10 times farther away, accompanied by its various smaller satellites and its spiral neighbor, M33 (see Chapter 9). At comparable but slightly lesser distances were the two small, irregular galaxies, NGC 6822 and IC 1613. Altogether, we found ourselves surrounded by a grouping of galaxies that appeared loosely clustered and isolated in space, with no other bright galaxies nearby. The nearest examples outside our local family seemed to be fairly remote, several million light-years farther out.

Fathoming the Local Group

As imaging technologies improved, astronomers identified more and more members of our Local Group (see Figure 8.1). Two distant companions to M31, the dwarf elliptical galaxies NGC 147 and NGC 185, were identified.

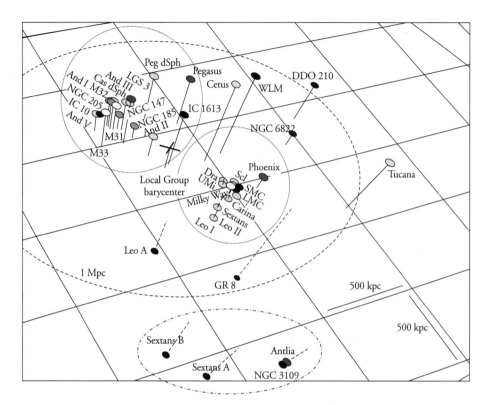

FIGURE 8.1 Mapping of the Local Group of galaxies, based on recent determinations of distance between the Milky Way and other galaxian members. The Local Group is dominated by the Milky Way and Andromeda (M31), with many of the dwarf galaxies bound as satellites to one of these two giants.

Other dwarf irregular objects were found that might be like NGC 6822 but appeared even less populous and smaller. Astronomers also found several very faint dwarf elliptical galaxies, starting in 1938 with the discovery of the Sculptor galaxy—a barely discernible haze on the photographic plate that first recorded the system. The population of our cluster gradually grew until by now astronomers identify about 40 objects in the Local Group, the vast majority being dwarfs.

How were these objects recognized as neighbors of the Milky Way? It was not always easy. For giant galaxies like M31, astronomers looked for Cepheid variables and then applied the period-luminosity relation to get the Cepheid luminosities, whose comparison with the apparent brightnesses yielded the distance. For other kinds of objects, however, this method did not always work. For example, the dwarf elliptical NGC 147 near M31 has no Cepheid variables. Because Cepheids represent a brief stage in the evolution of massive stars with short lifetimes, less than 100 million years or so, we can expect to find Cepheids only in galaxies where star formation has gone on in the last 100 million years. For NGC 147, star formation apparently ceased a few billion years ago. Thus there are now no normal Cepheids, and so other methods must be used to gauge its distance.

For this galaxy as well as its companion, NGC 185, and the two other small elliptical galaxies adjacent to M31 (NGC 205 and M32), the distance problem was solved by Walter Baade in 1944. Baade recognized that faint galaxies like NGC 147 were Population II objects, similar in stellar population to globular clusters in the Milky Way. Shapley had shown long before that the brightest stars in globular clusters all have very nearly the same intrinsic luminosities, and hence these stars could be used as standard candles to obtain reasonably good distances. Baade proceeded to use the brightest stars in the dwarf ellipticals as distance calibrators, assuming the brightest Population II stars all to be similar. This study led him to conclude that all four dwarf ellipticals near M31—NGC 147, NGC 185, NGC 205, and M32—must be members of the Local Group, and at about the same distance from us as M31. Several more faint companions to M31 have been discovered since Baade's pioneering work. They are described later on in this chapter.

The Local Group of galaxies becomes especially well defined when we examine the velocities of galaxies in nearby space. Beyond the Local Group, galaxies are receding from each other as they participate in the spectacular expansion of the Universe—set into play by the Big Bang. Inside our group, the motions are distinctly different. The combined masses of the members of the group exert enough gravitational attraction to counter the universal expansion, and so bind our family together in a cosy little corner of space.

Except for the occasional dwarf such as GR8 that lies just beyond the Local Group's boundaries, the surrounding space is pretty much devoid of galaxies. The Local Group spans about 5 million light-years, whereas the nearest other groups are 10 to 15 million light-years away. Thus we live in a small community of galaxies, surrounded by a relatively "rural" expanse that is thinly populated with itinerant dwarfs.

Dwarf Tales

Even a cursory glimpse of Figure 8.1 shows that the dwarfs outnumber the giants by about ten to one. If the rest of the Universe is like our Local Group, then it is inhabited mostly by these galaxian lightweights. A more careful ac-

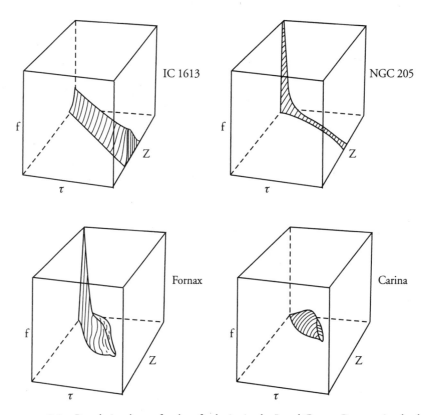

FIGURE 8.2   Population boxes for dwarf galaxies in the Local Group. By mapping both the star-formation rate (f) and chemical abundance (Z) as a function of age (τ), astronomers can trace a galaxy's evolutionary history. Here, the histories of the dwarf irregular IC 1613, the dwarf elliptical NGC 205, and the extremely faint Fornax and Carina dwarf ellipticals are compared. Remarkable variety is apparent in the patterns of star formation and chemical enrichment over cosmic time.

counting reveals tremendous variety in the types and luminosities of Local Group dwarfs, from the brightest dwarf irregulars to the faintest dwarf ellipticals. How these diverse systems came to be is of great interest to astronomers, for they provide critical clues to the evolution of all galaxies from the very earliest epochs to the present day.

As in the study of the Magellanic Clouds, population boxes help to visualize the sundry histories of star formation and chemical enrichment in the dwarfs (see Figure 8.2). Although all of the galaxies appear to have formed shortly after the putative Big Bang, some 15 billion years ago, the subsequent timelines of star formation and elemental enrichment have differed dramatically.

For example, the dwarf irregular galaxy IC 1613 has been producing stars at a steady clip throughout cosmic time, as evidenced by the equal numbers of old and young stars. Meanwhile, the initially low abundance of heavy elements has been slowly but steadily increasing to the modest level observed today in the Small Magellanic Cloud. By contrast, the dwarf elliptical galaxy Fornax started out forming stars like gangbusters, but petered out about 3 billion years ago. Its metallicity has held steady at SMC levels ever since. The extremely faint Carina dwarf elliptical presents one of the strangest cases, with its star formation peaking at an intermediate age of about 7 billion years (see Figure 8.2). Recent detailed observations of Carina has yielded even weirder results, with multiple epochs of star formation punctuating the history of this waifish dwarf.

Nowhere in these tiny realms is there the spread in chemical abundances that is evident in the giant Milky Way and Andromeda galaxies. Instead, each dwarf appears chemically pure: each collection of stars resembling a specially prepared petri dish of organisms that are co-evolving.

To further understand the nature of these unique realms, it is best to examine them one at a time. We will survey them here by type. First we discuss the irregularly shaped, fragment-like dwarfs and then we turn to the even less conspicuous class of dwarf elliptical galaxies, beginning with the dense companions to Andromeda, and ending with the ghostly systems that accompany both Andromeda and our own Galaxy.

Dwarf Irregulars

This class of motley-shaped galaxies constitutes nearly half of the dwarfs in the Local Group. Most of these irregular systems contain interstellar gas and show slow rotation, with speeds of 35–50 kilometers per second. In these regards, they are more like the Magellanic irregulars than their dwarf elliptical siblings.

NGC 6822

The brightest of the dwarf irregulars is NGC 6822, sometimes known as Barnard's galaxy (see Figure 8.3). Studied long ago by Hubble, NGC 6822 was the subject of his first paper on nearby galaxies, and in that sense can be considered the first recognized galaxy beyond the Milky Way. Recent work has extended Hubble's research. Newly measured Cepheids provide a new period-luminosity diagram and an improved distance estimate of 1.6 million

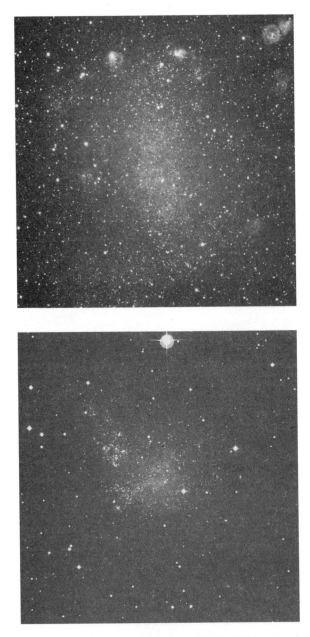

FIGURE 8.3   Comparison of the irregular galaxies NGC 6822 (top) and IC 1613 (bottom). These systems respectively resemble low-luminosity versions of the LMC and SMC.

light-years. Among the galaxy's brighter stars are numerous young, blue stars and a strong population of evolved giant stars.

The visually brightest stars are not the hottest, but have temperatures of about 20,000 K. They are not as luminous as the Magellanic Clouds' brightest stars, a probable result of the galaxy's overall smaller population. The red supergiants, however, have about the same luminosity as those in the Magellanic Clouds. That is because stars of a variety of masses tend to evolve into red supergiants of about the same maximum luminosity.

NGC 6822 has 16 bright stellar associations, averaging about 500 light-years in diameter. Star clusters are also present, with 30 clusters identified so far, ranging in age from a few million years to at least a billion years. The globular clusters of NGC 6822 tell an intriguing story. When Hubble published his paper on the galaxy in 1925, he claimed to have identified five objects that might be globular clusters. On his photographic plates they appeared as small, fuzzy objects whose brightnesses were about right for Galactic-type globular clusters. Little was done about these objects for many years, until in the 1960s astronomers measured their colors and concluded that Hubble was right, at least for four of them.

When the Hubble Space Telescope was trained on these small sources, however, a remarkable fact emerged. Only one of them, called Hubble VII, was truly like our globulars. It was measured to have an age of about 15 billion years, making it as old as the oldest stars we know. Another one, called Hubble VIII, was found to be younger, a little under 2 billion years old, and a third, Hubble VI, was a youthful 100 million years old. The fourth and fifth Hubble "clusters" were not even clusters, but were compact H II regions, gas clouds that Hubble mistook for unresolved groups of stars. They are not clusters, at least not yet, because they are still forming stars right now.

This spread in age is a very different situation from that in our Galaxy, where all of the globular clusters are about the same age, about 15 billion years. Apparently NGC 6822, like the LMC, has learned the trick of making globular clusters right up to the present, while our Galaxy stopped doing so long ago. This disparate behavior is one of the puzzles that astronomers are pursuing in their exploration of the nearest galaxies.

Among the youngest objects in a galaxy, the H II regions are often the most obvious. This is certainly true of NGC 6822, whose brightest gas clouds were the only things that some of its first observers could find. Hubble found 5, and subsequent searches have turned up many more; about 140 are now known, and there are probably many more faint ones. The biggest and brightest are beautifully structured complexes of gas and hot stars, smaller than 30 Doradus but not unlike it in structure. They are bright enough to allow detailed spectroscopic study. Using this diagnostic technique, astronomers at the University of Mexico have found that helium and other detectable heavier elements are lower in abundance in NGC 6822 than in our Galaxy and the Magellanic Clouds.

Neutral atomic hydrogen gas in NGC 6822 has been mapped with radio telescopes, though in less detail than the optically shining gas, mainly because of confusion with the foreground hydrogen in our own Galaxy. A huge envelope of thin, slowly revolving gas was found to extend far beyond the optically visible part of NGC 6822. Inside it is a bright, patchy arrangement of gas coinciding with the galaxy's optical core. Dense concentrations of gas occur primarily where bright OB stellar associations are located.

The rotation of the galaxy's core shows up clearly on the H I maps, the velocities varying smoothly from one end of the galaxy to the other, with a total kinematic range of about 75 kilometers per second. The dynamical mass derived from this study amounts to 1.6 billion Suns; this refers to all material inside a radius of 8.5 thousand light-years.

The optical structure of NGC 6822 is fairly simple, and was described qualitatively by Hubble. There is a vertical barlike core that is crossed by young objects (H II regions, stellar associations, and H I gas) in a T-shaped configuration (see Figure 8.3). Less conspicuous is a broad tail of material that swings off the east from the bottom of the bar. In gross structure, NGC 6822 is similar to the LMC; patches of active star-forming regions are scattered across a barlike assemblage of older stars, with the most prominent concentrated near one end of the bar.

Attempts to dig back into the history of star formation in both galaxies show that the same asymmetric pattern is prevalent as far back as can be traced. Age-dating the star clusters and stellar associations, we find that star-forming locales have bounced around from one place to another in a nearly random way in NGC 6822's recent history. The rate of cluster formation is quite low—on the average, one every 6 million years. The LMC, by comparison, manages to form a cluster every 30,000 years.

IC 1613

We have devoted several paragraphs to NGC 6822 not because it is a particularly noteworthy member of the Local Group but because it is a good example of a dwarf irregular galaxy that has been well studied. Similar work has been done on other dwarf irregulars in our neighborhood, most notably on IC 1613, a faint galaxy at about the same distance from us as NGC 6822 (see Figure 8.3).

Since NGC 6822 is structurally akin to the LMC, but smaller, it could be said that IC 1613 resembles a mini version of the SMC. Overall, it is 10 times fainter than the SMC. Its brightest stars are also fainter, its H II regions less massive, its star clusters, if any, almost invisible, and its Cepheids few in number. Otherwise, all of its observable components appear normal. The only abundant commodity in IC 1613 is the neutral hydrogen gas, which makes up almost 20 percent of the total mass in detectable form—a value about as large as is found in any normal galaxy yet studied. Hydrogen gas amounting to 70 million times the mass of the Sun is spread across IC 1613's pale face.

A long-standing puzzle has to do with star clusters in IC 1613. Where are they? Walter Baade first pointed out in the 1950s that in all the many years he had studied this galaxy, he had found no clusters at all. The galaxy's mass is about half that of NGC 6822 (about 800 million Suns), and yet there is no sign of any populous clusters like those discovered by Hubble in NGC 6822. Clearly, we are being told that some agent, yet unrecognized, governs whether or not a galaxy can form rich star clusters.

Because IC 1613 is a very small galaxy, it is expected to be deficient in heavy elements compared to the Milky Way. This expectation is borne out by the most recent spectroscopic data. IC 1613 is at least as low in these elements as NGC 6822, with a concomitant dearth of dust. Only 11 tiny dust clouds have been discovered, the smallest number in any galaxy that has been searched for these dark, obscuring objects. Observations of the far-infrared radiation that is given off by interstellar dust also shows that IC 1613 is nearly dust free. The Infrared Astronomy Satellite (IRAS), for instance, barely detected anything. Background galaxy counts also indicate only a very minor obscuring effect. At most, only 50 percent of faint background galaxies near the center of IC 1613 are screened by a general layer of dust. For the SMC, a comparable search shows an 85 percent decrease; therefore, there is proportionately less dust in IC 1613 than even in the dust-poor SMC.

Other Irregulars

There are several more dwarf irregular galaxies inhabiting our Local Group. A noteworthy example is the highly obscured galaxy IC 10. It lies beyond the opposite side of our Galaxy and is seen but dimly through the dust sheet of the Milky Way. Studies made near the beginning of the twentieth century showed that it is a special kind of galaxy, one of the hyperactive objects that astronomers call starburst galaxies (see Chapter 10). IC 10 has spread across its obscured face a splendid array of furious stellar furnaces, where stars are forming in great numbers. Quieter examples include quaint little objects like the Pegasus dwarf. This faint system may represent a transitional phase between the dwarf irregulars and the low-luminosity dwarf ellipticals discussed below.

Dwarf Ellipticals

The Local Group's dwarf ellipticals span an incredible range of luminosity and surface brightness—from M32, a dense starball resembling a downsized version of a giant elliptical, to Sculptor, a barely discernible mist of stars in the firmament. The Sculptor-type dwarf ellipticals are so faint and elusive that they are often classified separately as dwarf spheroidals, or dSph. They are not, however, necessarily spheroidal in shape, nor are they fundamentally different from the brighter dwarf ellipticals. Both kinds of galaxies have radial light profiles that fall off like those of their giant elliptical counterparts. That means they are all dynamically "relaxed" systems characterized by thermalized swarms of stars, in marked contrast to the rotating dwarf irregulars.

The obvious differences in appearance among the dwarf ellipticals amount to differences in overall luminosity (by as much as 10,000) and in central concentration. Whether or not the bright and faint dwarf ellipticals share similar pedigrees remains unknown. Indeed, the faint ellipticals may have evolved from dwarf irregulars whose gas had been previously stripped by internal or external means. According to this scenario, the remaining stars would have thermalized their motions over time, thus erasing any vestige of their rotating irregular forebears.

Companions to Andromeda

The Andromeda galaxy (M31), like the Milky Way, is surrounded by a family of dwarf galaxies (see Plate 16). Its nearest companions, rather than being two close irregular galaxies like the Magellanic Clouds, are two elliptical galaxies, M32 and NGC 205. From our vantage point, both are superposed onto the image of M31, with M32 appearing buried in the south-central outer spiral arms and the larger, but less prominent NGC 205 set among the faint outer stars of the main galaxy's northern regions. Two similar dwarf elliptical galaxies, only a little fainter than these, are found about 6 degrees away. They are NGC 147 and NGC 185, close enough to each other to make a probable binary pair. The rest of Andromeda's known companions, found between 1970 and 1998, are six very faint dwarf galaxies, usually

FIGURE 8.4   Companions of the Andromeda galaxy (M31). Each frame is 10,000 light-years on a side. Top left: M32 is an unusually bright dwarf elliptical whose concentrated starlight suggests that it was once the core of a much larger galaxy. Top right: NGC 205 contains some young stars mixed with its dominant population of ancient stars. Bottom left: NGC 185 contains clouds of dust and other tracers of continuing low-level star formation. Bottom right: NGC 147 is thought to be a faint binary companion to NGC 185.

called And Dwfs I, II, III, V, VII, and VII. The object named And IV is a faint irregular galaxy that was determined in 2001 to lie well beyond M31 and the Local Group, but it is seen through the outer folds of M31.

M32

M32 is a moderately bright, quite compact galaxy, with a total luminosity about that of the SMC, but with its light all packed into a nearly circular, smooth, bright image 6,000 light-years across (see Figure 8.4). It is visible in most pictures of M31 as the bright, nearly circular object south of the center (see Plate 16). M32 is primarily made up of intrinsically faint stars, mostly old stars somewhat like those of globular clusters. But its spectrum and red color indicate that its stars have chemical abundances different from those of the old, metal-poor globular clusters. Instead, there is a population of metal-rich stars, together with a component of intermediate-age stars (probably only 2 or 3 billion years old) making up a minor contaminant of the older population. Sparsely scattered among these stars are several planetary nebulae, but nothing else of a nebular nature. There are no great glowing gas clouds, no dust lanes, no neutral hydrogen gas—none of the things that a galaxy must have to form new stars. M32 is an elderly object with some middle-aged parts, but no signs of youth.

The high surface brightness of M32 has made it a favorite target for optical astronomers, especially spectroscopists, who can get data for it rapidly. Consequently, M32 has become a handy comparison object in many galaxy surveys. The mass estimate of M32 is derived from its range of observed velocities. Its spectral lines, which are widened by the Doppler shift, show that the full range of velocities near the center is about 140 kilometers per second. Thus some stars are moving away from the nucleus at 70 kilometers per second, while others are coming toward us at that velocity. From these velocities and the stars' distances from the center, we can calculate that the main body of M32 has a mass of about 6 billion solar masses, less than 1 percent of the mass of its "parent" galaxy, M31.

Even though its overall mass is small, M32 is apparently responsible for a considerable amount of disturbance evident within its host galaxy. M31's fragile spiral structure shows a confusing and complex, irregular pattern near M32. The arms of neutral hydrogen gas are displaced from the arms of stars by some 4,000 light-years. Moreover, it is impossible to trace the gaseous and stellar arms continuously through this disturbed area. Computer models of M31 show that a distorted pattern much like that seen can be produced by a close encounter between M31 and a companion galaxy with the mass of M32.

No doubt, Andromeda has had its revenge on M32. Multicolor data for the compact dwarf galaxy suggest that its stellar population is like that of a much more luminous elliptical galaxy. Because its compactness is also unusual for a galaxy of its luminosity, possibly M32 was once a much larger object, before tidal interactions with M31 stripped away its outer stars. Perhaps these stars have since mingled with the stars of M31, leaving only a dense core behind.

Indeed, an alarmingly high fraction of M32's overall mass appears to be contained in its unresolved nucleus. Velocity measurements there indicate the presence of a 3 million solar-mass black hole. Such a ponderous nucleus in such a small galaxy suggests that M32 may have once constituted the central core of a much larger elliptical or spiral galaxy.

NGC 205

The Andromeda galaxy's other close companion, NGC 205, has some peculiarities of its own (see Figure 8.4). On photographs it looks like an ordinary elliptical galaxy, rather elongated, of type dE5, indicating a dwarf elliptical with a ratio of short-to-long axes of about 0.5. It is not particularly luminous, and is only a little brighter than M32. Although twice the diameter of M32, NGC 205 is still 10 times smaller than M31. Both the shape and content of NGC 205 are known to be somewhat strange. Elliptical galaxies are normally symmetrical, with isophotes that are concentric, similar ellipses. In NGC 205, however, the very outer regions show a marked twist, with the major axis of the image making a fairly sudden change of direction.

This kind of peculiarity is not unheard of in elliptical galaxies; in fact, it is thought in other cases to be an optical effect that results from our viewing a somewhat prolate (elongated) galaxy from a particular angle. But for NGC 205, the twist is different; it shows up only in the very outer isophotes, rather than being a gradual shifting of the axis. In that way, it more nearly resembles the warping of the outer disks in our Galaxy and M31, where tidal effects from companion galaxies have probably altered the outer stellar orbits. The outer reaches of NGC 205, though not in a flat disk like that of a spiral galaxy, could also be responding to tidal forces. Thus there seems to be evidence that the galaxies in the Local Group are interacting with each other, causing distortions, tearing pieces of each other away, and producing peculiarities of structure that can help us to reconstruct their past.

The content of NGC 205 is its other anomalous property. Whereas elliptical galaxies are supposed to consist almost entirely of old stars, with little or no interstellar material, NGC 205 was shown by Walter Baade in 1944 to have a small population of very young stars, as well as some conspicuous dust lanes. Recent deep probes have turned up hot O and B stars, luminous carbon stars, and supergiant variable stars. Radio astronomers have also discovered a large thin cloud of neutral hydrogen gas in the galaxy. Thus NGC 205 has the raw materials to form new stars, as well as the inclination to do so.

NGC 185

The dwarf elliptical NGC 185 is noted for its unusual interstellar content. Although somewhat smaller and fainter than NGC 205, it is otherwise rather similar (see Figure 8.4). Instead of having a pure population of old stars, it too contains a small proportion of young stars along with some gas and dust. Two large dust clouds make up a patchy, nearly circular shell centered on the galaxy, about 200 light-years from its center. Young, blue stars are spread over the face of the galaxy, particularly near the center, whereas the neutral hydrogen is concentrated to one side.

A possible interpretation of this remarkable situation is that NGC 185 held back some of its gas at the time of most of its star formation, some 10 billion years ago. That gas is only now being used to form stars. The stellar winds from the concentration of stars near the center of the galaxy, though low, may be enough to explain the nearly circular shell of dust. As the starlight pushes against the grains of dust, it prevents the dust from falling to the gravitational center. The residual gas and the dust is small in amount (only a few million Suns in mass), but enough to form stars at a slow rate or, perhaps, in only a recent small burst. About 100 young stars are detected on deep photographic exposures of NGC 185.

It has been argued that the admixture of young stars is most likely a normal feature of elliptical galaxies. These stars are thought to be formed from left over primordial gas or from gas ejected by stars through normal stellar evolution, and are probably found in *all* elliptical galaxies. The spectra of many elliptical galaxies do suggest the faint presence of young, blue components, the relative importance of which increases with decreasing luminosity. Therefore, it seems plausible that there may always be a modest amount of star-forming activity, even in supposedly ancient galaxies like these ellipticals.

Andromeda's Faintest Companions

NGC 147 is even fainter than NGC 185 and is even more inconspicuous (see Figure 8.4). Like NGC 185, it has a few globular clusters, but no gas, dust, or young stars have been detected. Even though there is little evidence for recent star formation, the presence of carbon stars, red giant stars with strong bands of diatomic carbon (C_2) in their spectra, indicates that it is not exclusively made up of ancient stars. At least some star formation must have happened within the last billion years or so.

The galaxies known as And I, II, III, V, VI, and VII are very much like the nearer Sculptor-type galaxies in almost every way. Astronomers have measured the luminosities and colors of their brightest stars, a marvelous feat considering their faintness. These efforts have revealed a predominantly old, metal-poor population of giant stars.

The Nearest—and Faintest—Dwarf Ellipticals

The first two dwarf elliptical companions of the Milky Way to be discovered were reported by Shapley in 1938. They were named Sculptor and Fornax after the constellations in which they were found. Appearing as hazy smudges on the discovery plates, they were subsequently found to contain a smattering of old stars. These included a few RR Lyrae variables whose well-known intrinsic luminosities could be compared with their observed brightnesses for the purpose of deriving distances. The resulting distance estimates were about 260,000 light-years for Sculptor and 450,000 light-years for Fornax.

For the next 40 years, the Sculptor and Fornax galaxies remained examples of what was thought to be a pure globular-cluster population of stars, in the form of very spread-out, low-density galaxies. These systems have come to be classified as either dwarf elliptical (dE) or dwarf spheroidal (dSph) galaxies (see Figure 8.5). Seven more examples have been found in the Local Group. Four of them (Draco, Ursa Minor, Leo I, and Leo II) were discovered on photographic plates taken with the Palomar 48-inch telescope in the 1950s, when it was making its first deep survey of the northern skies. The Carina dwarf (1977), the Tucana dwarf (1992), and the Sagittarius dwarf (1994) were all discovered with the 48-inch survey telescope in southeastern Australia.

At first, these objects were regarded by most practitioners in the field as "merely" overblown globular clusters. In the 1980s, however, a surge of interest developed, primarily because several dissimilarities were found between the faint dwarf ellipticals and normal globular clusters.

The first major dissimilarity to be noticed was the presence of unusual variable stars, now referred to usually as anomalous Cepheids. They are like RR Lyrae variables in some ways, but have longer periods of pulsating luminosity (around 1 to 3 days) and are intrinsically brighter overall. One of the dwarfs, Leo I, has more than 50 of these anomalous Cepheids. The anomalous Cepheids have calculated masses that imply ages of only about a billion years. Therefore, they prove that the faint dwarf elliptical galaxies contain intermediate-age as well as old stars, unlike our Galaxy's globular clusters, whose stars are all old, with ages of 12–14 billion years.

Further evidence that the Sculptor-type galaxies have some young stars comes from the recent discovery of carbon stars in them. The luminous carbon stars, found in small numbers among the thousands of bright red giant

FIGURE 8.5   Faint dwarf companions of the Milky Way. Top left: The Phoenix galaxy represents a transitional object between the dwarf irregular galaxies and the even fainter dwarf spheroidals. Its interstellar medium may have been blown out by an episode of star formation as recent as a few hundred million years ago. Top right: The Leo II dwarf spheroidal galaxy contains predominantly old stars whose motions indicate the presence of gravitating dark matter. About 85 percent of this galaxy's mass is thought to be invisible. Bottom left: The Sculptor dwarf spheroidal galaxy was the first of its class to be discovered. Like Fornax, it is a barren system of ancient stars. Bottom right: The Carina dwarf spheroidal barely registers above the foreground of stars in our Galaxy. It is among the faintest and least substantial of the known galaxies. It is further noted for its strangely retarded and complex star-forming history.

stars in these galaxies, are thought to be fairly massive, with calculated ages (still somewhat tentative) of 1–2 billion years or so.

Color-magnitude diagrams of the Sculptor galaxies are also different from those of globular clusters in various ways. The red giant stars are spread out more in color and luminosity, apparently because of spreads in both age and chemical composition, and the bluer giant stars have unusual luminosities for their color. Furthermore, images obtained with the Hubble Space Telescope show the presence of intermediate-age main-sequence stars in some of the Sculptor-type dwarfs.

These telltale signs may be implicating the faint dwarf ellipticals as the skeletons of small *irregular* galaxies that finally exhausted their stellar-building materials (gas and dust) 2 or 3 billion years ago. Possibly the gas was swept out of them by collisions with the outer parts of the Milky Way, or by the violent blasts of their own supernovae. Since then, the remaining stars have dynamically "relaxed" into a thermalized swarm. This interpretation of the Sculptor-type dwarfs seems to explain their diffuse structure, which can be likened more to that of irregular galaxies than to that of normal elliptical galaxies.

One last curiosity about the Sculptor galaxies may be significant. The largest and intrinsically brightest ones, Fornax and Sagittarius, each have several globular clusters within and surrounding them. The globulars appear normal in just about every respect. The three that were discovered in Fornax at the time of Shapley's exploratory studies led to the conclusion among most astronomers that the Sculptor-type objects were not merely globular clusters that happened to have very low densities and large distances but were instead bona fide galaxies. Try to imagine galaxies as dogs and clusters as fleas. "Dogs have fleas, but fleas don't have fleas," would then be one way of putting it.

This flea-infested analogy is based on current models of hierarchical galaxy formation, where clumps of various size aggregated to form what we see today. In these models the globular clusters of our Galaxy are explained as very early condensations in the pre-Galactic gas cloud(s), which separated from other density fluctuations in the early Universe not long after the Big Bang. The protoclusters' densities were high enough for stars to form in them before much star formation proceeded elsewhere in the Galaxy, perhaps even before the disk of the Galaxy took shape. The globular clusters, in most models of this poorly understood process, need "host" galaxies to form nearby. Thus Fornax must be a true galaxy, as its six "fleas" testify.

Demise of a Dwarf

Before we leave the dwarf galaxies, we must visit one that has almost, but not quite, met its demise. Discovered only in the 1990s, the Sagittarius dwarf was previously invisible, lost amidst the myriad stars of the central regions of our Galaxy. When it was finally recognized, it turned out to be in a remarkable state. Spread out above the disk of the Milky Way, Sagittarius is slowly being devoured by our much larger and gravitationally dominant Galaxy. Before long, it will no longer exist, its stars and clusters having been added to the stars of the Milky Way. This process, called hierarchical merging, is fairly common in the Universe and probably happened more often in times long past—when the Universe was much denser. In fact, much of our Galaxy could have formed through mergers of smaller objects like the Sagittarius dwarf. How many little galaxies have given up their identities to build the Milky Way that we inhabit today? We don't know, but the number was probably at least in the hundreds.

9

T H E   N E A R E S T

G I A N T S

We may inhabit the surface of a minuscule planet that is in orbit about an unremarkable star. But the galaxy that holds our Sun and Solar System in its thrall is surely something to brag about. Home to several hundred billion suns and its own family of satellite dwarf galaxies, the Milky Way can be regarded as a bona fide giant. If the Universe of galaxies were like some sort of enormous game, we would do well by rooting for the home team.

In many ways, it is not so surprising that we find ourselves a part of this sprawling realm. Within its fecund disk, thousands of generations of massive stars have come and gone, their violent exhalations seeding the interstellar medium with the elemental stuff necessary to build up rocky planets. Oceans and atmospheres may coat a significant fraction of our Galaxy's rocky worlds, and in some of these moist domains carbonaceous life forms may be thriving.

Beyond the Milky Way are other giants, each one having its own grand evolutionary tale to tell. In this chapter, we will begin with the two other giants in the Local Group of galaxies, the nearest giant spiral, M31, and its less massive but spunkier neighbor, M33. We must then extend our view beyond the Local Group to explore the nearest examples of giant spirals that feature prominent arms, central bars, and starbursting rings. Our tour ends with the pathological giant elliptical Centaurus A. There are many more giants to explore, of course, with several dozen within easy reach of small amateur telescopes.

M31, the Great Nebula in Andromeda

Among the many objects in the northern sky visible through small telescopes, M31 is one of the most special (see Plate 16). It is immense, spanning more than a degree in the sky—the equivalent of more than two Moons laid end to end. It outshines most other nebulae in the night sky, and yet it is one of the most distant objects that can be seen with the unaided eye. Indeed, M31 is more than 1,000 times farther than most other visible nebulae, lying in the remote reaches of the Local Group some 2.5 million light-years away. But its most spectacular feature is its sheer beauty, whether it is observed faintly through binoculars or admired in all its glowing glory through a big telescope. Photographs reveal its amazing structural variety, from its luminous, yellow central bulge to its enfolding spiral arms adorned with blue stars, clusters, and dark lanes of dust. As an example of a giant spiral galaxy, M31 is especially well endowed with the sorts of things that astronomers must study up close before hoping to understand the galaxies far off in the distant Universe.

Hubble's epochal study of M31, published in 1929, first proved the true nature of this nebula as a galaxy. After measuring M31's distance using the Cepheid variable stars that he had discovered in its spiral arms, Hubble proceeded to discover many other features which confirmed his view that the Great Nebula must be a galaxy. In particular, by comparing many of the photographs he had taken over the years, he found 63 novae—stars that suddenly light up to great brilliance and then, in a matter of days to months, fade away to invisibility. The novae of M31 became as bright as the most luminous Cepheids, some even brighter, telling Hubble that they must be similar to the novae of our Galaxy. Hubble did not know the cause of novae, but he knew how intrinsically bright they get in the Milky Way, and so could conclude that M31's novae confirmed the distance he had derived from the Cepheids.

Nova Pyrotechnics

We now understand more about what causes a nova. Each nova is thought to arise from a close binary star system, with one member of the pair being a collapsed star known as a white dwarf. Having completed its normal life, a star of intermediate mass is reduced to a faint, very dense, hot "cinder," with a mass similar to that of the Sun but a size closer to that of Earth. Because of the white dwarf's extremely high surface gravity, anything that happens to fall onto its surface will set off a reaction like an atomic bomb. Indeed, the nova outburst is attributed to just such a reaction, known as a thermonuclear runaway.

The other star in a nova system is an evolving star that has grown in size as it enters its red giant phase. It is lagging behind its companion in its evolutionary sequence, probably because it is a little less massive. As it grows in diameter, some of its outer gas layers feel the pull of the denser dwarf companion. The hydrogen-rich gas ultimately streams across the gap to fall onto the companion's hot surface. Compressed by the dwarf's huge gravity, the hydrogen fuses into helium, releasing enough energy in the thermonuclear runaway to form a brief, spectacular nova outburst.

Almost 20 years after Hubble's paper was published, a more systematic and complete study of the novae in M31 was undertaken at Mount Wilson. For an entire year the galaxy was photographed on every available clear night. The result was the first study of its kind, a complete year-long record of the novae in a galaxy. Nothing like it is possible for the Milky Way, because dust hides most of the Galactic novae from our optical view. Other galaxies are either too small to have frequent novae or else too far away for their novae to be readily observed. The nova survey, like so many of the studies of M31, is important not only for understanding that galaxy but also for learning new things about galaxies in general—insights that we could not obtain otherwise.

For instance, an interesting and important feature of the properties of novae turned up, something that would not have been so easily discovered in our local Galaxy. Different novae were found to decay in brightness at different rates. The rate of decay correlated with the maximum brightness of the novae, with the brightest novae decaying the most rapidly. All of the novae in M31 are at relatively the same distance from us, about 2.5 million light-years, so we can compare them directly without having to worry about the effects of different distances, as is the case for novae in our Galaxy. This important correlation shows up most clearly for the novae in M31, owing to the large number that have been observed there. If it is assumed that all novae everywhere behave similarly, then we can use the results from the M31 novae to gauge the luminosities of novae anywhere—provided we measure their rates of decay—and so surmise their distances. The novae of nearby galaxies, through this correlation, have become an important means for attacking the distance scale problem, which is discussed more thoroughly in Chapter 12.

S Andromeda, M31's Lone Supernova

When the first novae in M31 were being discovered at Mount Wilson in the early 1920s, they aroused some skepticism because of their faintness. Since the true distance to Andromeda was not then known, the faintness had no explanation, and the problem was compounded by another nova, one that had appeared a quarter of a century earlier. An extremely bright "star," almost visible without a telescope, had suddenly lit up in 1885 near the center of M31. Named S Andromeda, it behaved like a nova, gradually fading until a few months later it disappeared from view. Astronomers of the time assumed that it was an ordinary nova. During the 1920s, when the nature of galaxies was being debated, some astronomers pointed to the remarkable brightness of S Andromeda as evidence that M31 cannot be very far away. But when Hubble established the galaxy's distance on the basis of Cepheid variables, astronomers realized that S Andromeda could not be a true nova but must be another type of object that is a million times more powerful, what we now call a supernova.

Several of these whopping explosions had been seen long before in our own Galaxy, though not recognized as supernovae, while others had been observed occasionally in more distant spirals, again without proper recognition. We now know that supernovae involve the complete destruction of a massive star at the end of its lifetime, or of a white dwarf that has accreted too much mass from its red giant companion. The end result is an explosion as powerful as 3 billion Suns, all from one star. What remains is a rapidly fading nebula (see Plate 11), containing a neutron star the size of a typical city, a black hole of similar dimension, or maybe nothing at all. In a giant spiral like M31, supernovae are expected to occur every 50 years or so. S Andromeda is the only one seen in M31 thus far, so we should expect to witness another of these spectacular events very soon.

M31's Cepheid Variables

The Cepheid variables of M31 have been of great historical importance since Hubble detected 40 of these giant, pulsating stars in the 1920s and used them to gauge the galaxy's distance. About 25 years later Walter Baade and Henrietta Swope used the Palomar 200-inch telescope to make a more thorough study. Their photographic plates revealed hundreds of Cepheids. By carefully tracking the light from these stars, they were able to further refine the Cepheid period-luminosity relation, the single most important tool for establishing extragalactic distances.

More recent surveys have shown that M31 contains thousands of Cepheids. Most of these supergiant stars are located in the blue spiral arms, some 25,000–50,000 light-years from the nucleus. The observed properties of M31's Cepheids are very similar to those of Cepheids in our Galaxy, making them especially useful for establishing the first "rung" on the extragalactic distance "ladder" (see Chapter 12).

Matter between the Stars

There is gas in Andromeda, some of it visible but most not. The visible gas appears as relatively small and dim H II regions, in or near the galaxy's spiral arms. More than 1,000 of them are known, the result of surveys carried out by several groups of astronomers. The H II regions in M31 are less spectacular than those in the Magellanic Clouds. This deficiency is probably due to the relative paucity of massive star-forming regions in M31. Apparently, the gaseous and dynamical conditions in this swirling giant galaxy are less favorable for building and maintaining giant clouds than those found in the more serene and gas-rich Magellanic Clouds. Other Sb-type galaxies seem to be equally short-changed, so the galaxy's structure must somehow be involved in the explanation. Sa-type galaxies are even worse off; $H\alpha$ emission-line surveys turn up only a handful of Sa's that show many H II regions. The lower molecular gas content remaining and high rates of shearing in these bulge-dominated systems are at least partly to blame.

Cool, neutral hydrogen gas (H I), by contrast, is fairly abundant in M31, lying undisturbed in the vast voids between the stars. The neutral hydrogen in M31 has been studied in great detail with large radio telescope arrays, including the Westerbork array in Holland and the Very Large Array in New Mexico. The resulting maps show exquisite amounts of structure at every velocity, allowing us to trace the complex motions as well as the overall architecture of the gas. The neutral hydrogen is not spread out in the galaxy like the stars, but shows a large gap at the middle and a ringlike concentration at a radial distance of 30,000–40,000 light-years from the nucleus (see Figure 9.1). This ring, or pseudo-ring, is also where the visible gas clouds are most numerous and where the spiral structure is the most prominent. Unlike the optical H II regions, which taper off at radial distances of about 50,000 light-years, the neutral gas as traced in radio maps extends to a radius of 100,000 light-years, far beyond most of the visible parts of the galaxy. This is three times as far as the Sun is from the center of our Galaxy and 50 percent farther out than the most remote detected gas in the Milky Way's plane.

The hydrogen gas does not all swing around the center of M31 in perfect

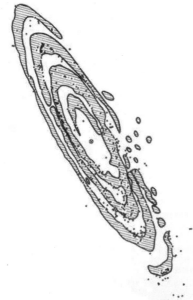

FIGURE 9.1 Gas and stars in the Andromeda galaxy (M31). Top: The neutral hydrogen arms (hatching) and open star clusters (dots) in M31 are closely associated. Bottom: The northeast part of the Andromeda galaxy (M31) features clusters and larger associations of hot young stars.

circles, as might be expected. Instead, some of it behaves quite strangely. The inner arm in the northeast part of M31 is falling in toward the center, as well as racing around in its orbit. The infall velocity is as large as 100 kilometers per second (about 200,000 miles per hour)! We still do not know the reason for this anomalous motion; it may be due to a recent merger event, or perhaps it results from tidal action on the arm caused by one of the existing companion galaxies.

Maps of radio continuum radiation emitted by M31 have been made with several of the world's biggest radio telescopes. A large amount of the radio light comes from the center of the galaxy, where a very bright source is located, probably a nonthermal source related to the nuclear activity of M31. Perhaps there is the remnant of a large collapsed object at the nucleus, probably a black hole, although M31 is certainly not a radio galaxy in the usual sense of the word (see Chapter 11). Other nonthermal radio light comes weakly from the disk of the galaxy. The intensity and character of the radio radiation is about what we would expect from the number of supernova remnants that should be there, based on a sample of objects like the Crab Nebula from our own Galaxy. A few of the remnants are just bright enough to be seen and analyzed optically. The radio and optical data indicate that the supernova remnants are concentrated near the central bulge and extend thinly throughout the disk.

The rest of the radio light that is detected seems to come from small H II regions in the disk and especially from the brighter parts of the spiral arms. There is a strong maximum at a distance of about 30,000 light-years, where the neutral hydrogen and star formation are also concentrated. A doughnut-shaped ring contains most of the action in M31, out to about the distance from the nucleus that corresponds to the Sun's distance from our Galaxy's nucleus.

Planetary nebulae are the expelled outer atmospheres of dying stars that are being photo-ionized by the hot stellar cores. They abound in the areas of M31 that are dominated by old stars—especially in the central, giant-rich bulge. Astronomers using the 3-meter telescope at Lick Observatory near San Jose, California, have discovered 315 planetary nebulae in just a small region of M31. They estimate that the whole galaxy has about 10,000 of these tiny, fluorescing objects.

Star Clusters and Associations

The open star clusters in Andromeda are nearly as small and hard to detect as the planetary nebulae (see Figure 9.1). Hubble most likely found a few because he pointed out an example in a photograph in one of his papers. But little was done about them until the 1980s, when the 4-meter telescope at Kitt Peak National Observatory near Tucson, Arizona, was trained on M31, and astronomers discovered more than 400 of these faint, indistinct objects. Most of the detected clusters were large and young, like the Double Cluster in Perseus (h and chi Persei), with ages of less than about 100 million years. Deeper surveys, made with the Hubble Space Telescope, have since revealed a lot more clusters. These results indicate that the whole galaxy probably has a few thousand open clusters of various ages. Such a wealth of stellar "families," once fully explored, will be especially useful for delineating the history of M31.

What M31 lacks in giant H II regions and bright star clusters it more than compensates for with its gigantic stellar associations. The 200 or so stellar associations that punctuate M31's spiral arms are surprisingly different from those found in the Milky Way. They contain the same kinds of brilliant blue stars and gas clouds, but are almost 10 times larger than the local sample. Instead of being about 150 light-years across, like the well-known associations in Orion and Sagittarius in our Galaxy, those in M31 are about 1,500 light-years across. We simply do not know why this difference occurs; other galaxies, such as the Magellanic Clouds, NGC 6822, and IC 1613, have stellar associations very similar in size to ours.

The presence of these stellar behemoths also begs the question of their origin. Where are the giant molecular clouds that could spawn such large assemblages? The low levels of CO emission that have been detected suggest that the molecular cloud component is rather meager. This molecular paucity further prompts one to consider the possibility that the abundant atomic clouds are in some way directly contributing to the process of building M31's associations. Perhaps instead, dense star-forming clouds of molecular gas are roaming among the atomic clouds, but happen to be in a state of low CO emissivity. Detections notwithstanding, recent mergers of gas-rich dwarf galaxies—or less mortal encounters with M31's sundry dwarf companions—may have helped to build up the giant associations that are observed today.

Globular clusters in M31 are easy to see, some even with moderate-sized telescopes (the brightest clusters are about magnitude 14). Hubble first catalogued nearly 200 of these luminous but unresolved objects; more recent efforts by independent teams of astronomers in California, Canada, Italy, and the former Soviet Union have tripled Hubble's number of globulars.

Recent observations by the Hubble Space Telescope easily resolve the clusters into their myriad stars. These observations are now being used as probes of the galaxy's old population, of its dynamical and chemical history, and of its total mass.

Andromeda was the first galaxy ever to have its mass estimated. As discussed in Chapter 4, the speed at which a star or gas cloud moves in its orbit around a galaxy depends upon the amount of mass lying inside its orbit and on the size of the orbit. By plotting out the velocity curve, an astronomer can measure the mass distribution in the galaxy and then sum it up to get the total mass.

For M31, the optical data, beginning with a landmark study in 1939, seemed to indicate that the mass must be about twice the mass that was then estimated for the Milky Way, roughly 200 billion times the mass of the Sun. But later radio observations disagreed with that conclusion. Most of M31 is way out where we can see nothing, far beyond the optical image and even beyond the radio limits. The revised estimate for the mass of M31 is now at more than a trillion Suns. Other galaxies, even our own, are similarly dominated by unaccountable dark matter. M31 was the first galaxy to present us with this cosmic puzzle.

Spiral and Ring Structures

The spiral arms of Andromeda are yet another enigma. Because we see the galaxy's disk to within 15 degrees of being edge-on, the different parts of its structure defy straightforward delineation. The structure is especially difficult to discern because there also seems to be a disturbance of the arm pattern due to the tidal effects of M32. Nevertheless, it is possible to perceive a two-armed spiral pattern with arms arranged so that they trail behind the rotation of the galaxy. Theoretical models of galaxies seem to indicate that this kind of rotation, with the arms trailing, is the most common arrangement.

A mapping of M31's far-infrared emission by the Infrared Space Observatory (ISO) has revealed a more ringlike distribution (see Figure 9.2). Rectification of this mapping to a face-on perspective makes the ring pattern even clearer. Because the far-infrared emission arises from dust and organic molecules that have been warmed by neighboring hot stars, it traces key components of the star-forming process. Therefore, the ringlike arrangement seen in M31's dust and gas provides a fascinating picture of the recent starbirth activity taking place in the galaxy's disk.

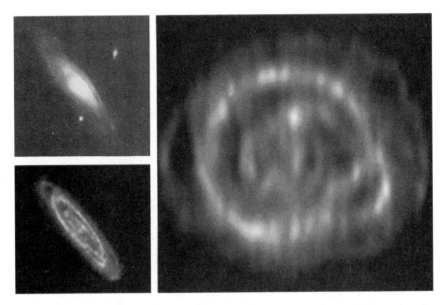

FIGURE 9.2   Andromeda's ringlike pattern. Top left: Andromeda at visible wavelengths shows a central bulge and disk at high inclination. Bottom left: The Infrared Space Observatory's map of the far-infrared emission from M31. Right: Rectification of the galaxy to a face-on configuration shows a distinct ring morphology. The far-infrared emission from the ring is tracing dust grains and organic molecules associated with regions of recent star formation.

Nuclear Duplicity

As the nearest giant spiral galaxy to us, M31 provides a critical benchmark for understanding galaxy structure and energetics. If we can find something in M31, we are more likely to expect similar things occurring in other, more distant giant spirals. So it is somewhat disconcerting to discover that M31 has a very strange nucleus. Besides radio and X-ray evidence for an actively feeding super-massive black hole, we now have optical evidence that M31's nucleus contains multiple components.

Using the Hubble Space Telescope, astronomers have found two sources in the core of M31. The brighter source looks like a giant globular cluster, while the other appears more like an extended disk with a compact UV-emitting star cluster superposed. The two sources are separated by only 0.5 arcsecond on the sky, or 5.5 light-years at the distance of M31. The fainter of the two sources appears to lie near the dynamical center of the galaxy, and so is thought to be the more massive and ancient nucleus. The motions of these two nuclear components suggest that both may be interacting gravitationally with an even more massive object in the core. The nature of this object is unknown, for it is completely dark. Indeed, a black hole with a mass of 33 million Suns provides one of the best explanations of the observed motions.

M33, the Pinwheel in Triangulum

About 15 degrees away from M31 lies another spiral galaxy, in the tiny constellation of Triangulum (see Plate 17). An easy object to see with small telescopes, M33 is bright enough to be barely visible to the unaided eye under near-perfect skies. It is both smaller and fainter than the Andromeda nebula, and because it is at about the same distance from us, these differences are intrinsic as well as apparent.

Through binoculars or a small telescope, M33 looks like a diffuse patch, whereas M31 shows a strong central concentration of light. Consequently, M33 is of a different Hubble type from M31, being a bulgeless example of type Sc, while M31 is Sb. It is a different luminosity-class object, too; its class is III, indicating its relative faintness. Compared with M31 and the Milky Way, M33 does not really qualify as a bona fide giant, but rather provides a transition between the Magellanic irregulars and the giant spirals. Fortunately for us, it is only moderately tilted to the plane of the sky, so that its spiral arms, fluorescent gas clouds, and many brilliant stars are nicely laid out for us to see and study.

Giant H II Regions and Associations

The H II regions of M33 are far more luminous and interesting than those of its larger neighbor, M31. Like the Magellanic Clouds, M33 has a number of true giants among its gas clouds; three are large and bright enough to rival the great 30 Doradus in the LMC. NGC 604, the biggest and brightest, contains a nest of furiously burning O stars, with temperatures as high as 50,000 K (see Figure 9.3). Some are Wolf-Rayet stars, whose hot outer atmospheres boil away in the intense heat and light, producing broad emission lines that make them easy to recognize even in the glare of the other stellar powerhouses. These very massive, young stars serve as primary indicators of active star formation.

Images taken with the Hubble Space Telescope of NGC 604 and the other giant H II regions in M33 cleanly resolve the underlying star clusters into their individual stellar constituents. Somewhat surprisingly, the stars appear far more loosely distributed than those crammed into the R136 cluster at the

FIGURE 9.3   M33, a relatively low-luminosity Sc galaxy in the Local Group, has a ragged spiral structure (left). The bright patch to the upper left (northeast) of the galaxy is the supergiant H II region NGC 604. Details of this stellar and nebular hotbed have been revealed by the Hubble Space Telescope (right).

core of 30 Doradus (see Figure 7.4). What these starbursting clusters lack in central concentration, they make up for in total real estate. Spanning some 300 light-years, these caches of newborn hot stars provide astronomers with ideal laboratories for investigating the effects of environment on the process of star formation.

For example, the giant H II regions closest to the nucleus are 1.4 times richer in heavy elements than what is observed in the Sun and the Orion Nebula, while those in the outer reaches of the galaxy have less than one-fifth the metallicity evident in the Solar Neighborhood. These elemental differences may have an effect on the types of star clusters that are created. Preliminary evidence indicates that such an effect may be present, with greater numbers of high-mass stars existing in the chemically rich H II regions. Because of its proximity, wealth of star-forming regions, and varying chemistry, M33 will continue to play a pivotal role as we try to understand the more powerful starburst activity occurring in distant galaxies (see Chapter 10).

The stellar associations of M33 have also come under recent scrutiny. There are 143 of them listed in current catalogues, and many of the galaxy's most luminous stars are located within them. Spectroscopic analysis of the brightest known star in M33, located in one of these associations, yields an absolute magnitude of −9.4, not quite as bright as the brightest supergiant stars in the Magellanic Clouds. Some very bright stars also occur outside the associations, despite astronomers' theoretical models, which say that it is almost impossible for massive stars to form in isolation. Much is still to be learned about the factors that contribute to star formation, especially the formation of large, massive stars between the spiral arms.

A Raggedy Spiral

The spiral structure of M33 is quite typical of low-luminosity Sc galaxies. There are two main arms, but they cannot be traced very far around the galaxy before they lose their distinctness (see Plate 17). The arms are fragmented into about 10 pieces, with the galaxy's stellar associations, H II regions, supergiant stars, and underlying unresolved stars all concentrated in them. The dust and neutral hydrogen gas components, on the other hand, are chaotic in their distribution and hence could not be used alone to discover the spiral structure.

This is a strange result: What we know about spiral structure tells us that dust and gas will accumulate in the crest of a spiral density wave, usually along the inside of the spiral arm. The dust is needed there, in order to help with the difficult task of spawning new stars. The subsequent star-forming effects are traced by the young hot stars and fluorescing gas clouds "downstream" from the dust and gas—that is, along the outer edge of the spiral arms. In some galaxies, at least, the dust is well behaved and lies conveniently along the inner side of the luminous spiral arms, but not in M33. It may be some time before this interesting mystery is solved, because the dust and associated gas in a galaxy are not very easy to study in the kind of detail astronomers would like.

The stochastic theory of spiral structure—through self-propagating star formation in the presence of differential rotation—does a somewhat better job of explaining the relatively disorganized structure observed in the dust and gas. According to this theory, luminous starbirth regions are generated randomly in the galaxy's chaotic and swirling interstellar medium. These regions are then sheared by the galaxy's differential rotation into spiral segments that eventually fade away—to be replaced by new starbirthing sites. If the stochastic theory applies to M33, then the spiral segments should show evidence of self-propagating star formation having occurred in them. The youngest and hottest stars should be found preferentially at one end of the segment, with a spatiotemporal sequence of more evolved stellar populations being found further along the segment. Such sequences should be evident in existing optical and ultraviolet images of M33, provided the spatial resolution is up to the task.

A Gaseous Froth

The neutral hydrogen gas in M33 has been mapped and found to be rather well behaved. It spreads over the whole face of the galaxy, terminating a bit beyond the visible disk. The velocities are fairly regular, indicating that most of the gas rotates around the nucleus in circular orbits. The total gravitating mass inside the outermost measured point is about 40 billion Suns, quite a bit less than the dynamical masses measured in M31 and the Milky Way.

Like the dust, the H I gas in M33 does not show much spiral structure. Instead, it displays a myriad of shell-like features, akin to those seen in the LMC. The origin of this frothy structure has yet to be elucidated, but at least some of the features are probably true bubbles inflated by prior starbirth and stardeath activity inside them. Kinematic mappings of these H I bubbles has shown that some are expanding at rates of 25–100 kilometers per second, and so have energetics requiring the equivalent of multiple supernova explosions. The supernova remnants have long since vanished, but the remaining stars that formed with those that exploded could reveal how long ago the outbursts occurred—if these stars can be located amongst the rich stellar background in the disk.

Open and Globular Clusters

The star clusters of M33 have long intrigued astronomers. Hubble found a number of objects that he thought might be globular clusters, but he noted that they were significantly fainter than those in M31. More recent work shows that they also have disparate colors. Most of Hubble's original 15 globular cluster candidates are blue, similar in color to relatively young open clusters like the Pleiades in our Galaxy. Evidently, they are actually just rich open clusters; the brightest need only be two or three times as bright as the Pleiades, which is not a particularly big cluster. The majority of the objects, then, are youngish clusters, like the blue globulars of the Magellanic Clouds, but larger and less concentrated. Recent, more complete surveys have isolated several hundred candidates, the majority of which are open clusters. Only a few could be genuine globulars, and these are poor examples of their class, faint and small.

Why M33 should be so poor in true globular clusters, while M31 is so rich, is an important question. We can speculate that it has something to do with M33's smaller mass, but it also must be a result of its different Hubble type. Apparently globular clusters do not form in such abundance or in such large size in galaxies of type Sc, Sd, Sm, or Im. This seems consistent with the fact that such galaxies also do not contain big central bulges of old stars. They apparently did not indulge in as much early star formation when they were first condensing out of the cosmic maelstrom—when bulge and globular cluster formation would have been most active in them.

An Active Nucleus?

The most recent surprise about M33 came to light only in 1983. The orbiting Einstein X-ray Observatory detected an X-ray source at the nucleus of M33. Other sources had been found in the disk, most of them being identified as supernova remnants or closely interacting binary systems involving neutron-star companions. The X-ray source at the nucleus seemed special, however, being similar to the X-ray sources associated with active galactic nuclei. Active galactic nuclei (AGN) inhabit the dense cores of the bizarre Seyfert galaxies and the blazing quasars (see Chapter 11). The putative AGN in M33 is nowhere as powerful as those, being a very small example of the class, though ten times as powerful as the object inhabiting the nucleus of M31 and over 10,000 times more powerful than the source at the nucleus of our own Galaxy.

An unusual feature of M33 X-8, as the source is called, is the fact that it does not show up at any other wavelength. Many AGN are strong emitters at all wavelengths of light, from gamma-ray to radio wavelengths. But M33's strange nuclear feature, perhaps a massive black hole or other collapsed object, emits all of its power as X-rays. At these wavelengths it shines a million times more brilliantly than the Sun does at all wavelengths. Recent observations with the Chandra X-ray Observatory may ultimately resolve the puzzling nature of M33's nuclear X-ray source.

M81, the Nearest Grand-Design Spiral

To sample the incredible variety of giant galaxies, one must leave the Local Group and investigate other nearby groups of galaxies. Located roughly 10 million light-years away in the constellation of Ursa Major, M81 (NGC 3031) is the closest example of a grand-design spiral galaxy (see Plate 18). This Sb-type galaxy is barely visible through binoculars, but shows a distinct bulge-plus-disk form through amateur telescopes with apertures 8 inches or more in diameter.

At optical wavelengths, M81 shows two broad spiral arms extending beyond its prominent bulge. At ultraviolet wavelengths, the bulge virtually disappears, and the spiral arms are reduced to thin trails of light (see Plate 4). There we are witnessing the hottest and youngest stars—recently forged in the spiral arms. If spiral density waves are responsible for having organized gas clouds into dense spiral configurations, then the UV-bright arms are delineating the stellar consequences—just downstream from the spiral density wave's crest.

The radio-emitting neutral hydrogen gas has been mapped with sufficient resolution to show the spiral-shaped accumulations (see Figure 9.4). From the Doppler shifts of the 21-centimeter- wavelength H I line emission, a map of the corresponding radial velocities can be generated. As shown in Figure 9.4, the gas velocities along the spiral arms are perturbed relative to the interarm velocities. These sorts of perturbations indicate streaming motions along the arms, just as predicted by the spiral density-wave theory.

When orbiting gas clouds approach a spiral density-wave crest, they feel the extra gravitation from the density wave and are pulled to a greater orbital radius. Conservation of angular momentum dictates that the clouds slow down at the greater orbital radius. Upon passing through the crest, the orbiting clouds are pulled by the gravitating wave crest to a smaller radius and faster orbital speed, eventually regaining their inter-arm trajectory. The resulting motions are observed as oppositely directed "streaming" velocities along the inside and outside of the arms.

In the case of M81, the streaming motions amount to about 50 kilometers per second, or about 25 percent of the orbital velocities at these radii. The northeast arm shows this streaming best, with the southwest arm showing a more complicated pattern of velocity perturbations. This difference may have something to do with the extremely busy environment in which M81 finds itself. Closely accompanying this giant spiral is the Magellanic irregular galaxy Holmberg IX, the amorphous galaxy NGC 3077, and the starbursting galaxy M82 (see Chapter 10). Moreover, all of these systems are awash in a vast pool of neutral hydrogen. The interactions between M81 and these various components may be responsible for both generating and perturbing the spiral density waves that ripple through this grand-design spiral.

One peculiar aspect of M81's grand design is the absence of spiral structure within a galactocentric radius of 16,000 light-years. The star-forming gas may be partly responsible, for it is also lacking within this regime. But the older stars of the inner disk should also show a broad spiral pattern in the

FIGURE 9.4 Neutral hydrogen gas in the "grand-design" spiral galaxy M81, as mapped with the Very Large Array (VLA) of radio telescopes. Top: The H I gas is configured in a nearly ideal logarithmic spiral. Bottom: A contour map (spider diagram) of the radial velocities in the H I gas indicates material rotating away from us at the upper right (northwest) and toward us in the lower left (southeast) of the galaxy. The numbers indicate the radial velocities in kilometers per second. Wiggles in the contours indicate streaming motions along the spiral arms.

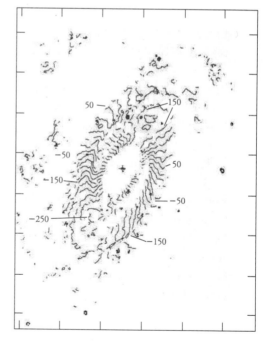

presence of density waves, which they do not. Somehow, the spiral density waves are impeded from penetrating into the inner disk.

One way to deflect the waves is to set up resonant conditions between the radially oscillating stars and their periodic encounters with the spiral wave-fronts, thus establishing a reflective "mirror" of sorts at the resonant radius (see the discussion of M94 below). In M81, there does seem to be a ringlike pattern of recent star formation at the threshold radius. But the necessary conditions for an orbital resonance seem to be absent in this part of the galaxy. Another way to thwart spiral structuring is to "heat up" the stars in the inner disk by increasing their relative motions, so that they no longer respond to the wave perturbations in any coherent manner. This seems to be the case in M81, where the stars of the inner disk show large velocity dispersions—perhaps due to the coexisting bulge and its gravitational effects on the disk stars.

M74, a Multi-Mode Grand-Design Spiral

Other notable grand-design spirals within easy reach include the Sc-type galaxy M74 (NGC 628) in Pisces and the barred SBc-type galaxy M83 (NGC 5236) in Hydra. Compared to M81, M74 shows spiral structure all the way into the nucleus (see Plate 18). In this nearly face-on system, spiral density waves most likely dominate the dynamics of the disk, with the small bulge being of negligible consequence. M74's spiral arms, however, appear disjointed at particular locations.

To account for these kinks in the spiral structure, theorists have considered outward- and inward-propagating density waves that lead to quasi-stationary modulations akin to the standing waves on a guitar string or in an organ pipe. Other theorists have added a set of three-armed density waves to the standard two-armed mode. This multi-mode approach, when properly tuned, yields interfering density waves that successfully account for the kinks as well as the periodic variations of brightness that are observed along the spiral arms.

M83, a Barred Grand-Design Spiral

In M83, the large central bar limits the radial extent of the spiral structure (see Plate 19). As in other barred grand-design spirals, the arms begin at the ends of the bar and then trail off for close to one full winding. The arms in M83 display nearly ideal offsets between the star-forming material and the luminous consequences, as predicted by density-wave theory. High-resolution radio maps place the CO-emitting molecular gas along the inner edges of the spiral arms—coincident with the silhouetting dust lanes. The Hα-emitting giant H II regions and associated O-type stars blaze forth along the arms' midsections, while the ultraviolet-emitting B-type stars and H I gas straggle "downstream" along the outer edges of the trailing arms. The H I gas in the arms appears to be a byproduct of the hot newborn stars, whose winds have disrupted the molecular gas clouds, and whose UV radiation has dissociated the molecular hydrogen back to its atomic state.

The bar itself provides the most likely generator and sustainer of the spiral density waves. Like any dissipative wave, the density waves energize the medium in which they propagate, and in so doing, they eventually peter out. The gravitating asymmetry of the bar is thought to reinvigorate the density waves that propagate through the disk, while simultaneously establishing radial flows of gas toward and away from the nucleus. In M83, the bar-induced flows help to explain the concentration of brilliant starburst activity that is observed within 1,000 light-years of the galaxy's nucleus.

The fact that M83's bar is undetectable at ultraviolet wavelengths immediately tells us that the bar is not a site of recent star formation—despite the abundance of gas along its length. Perhaps the streaming flows of gas in the bar are too disruptive for cloud and star formation to take hold. Or perhaps the residence time of gas in the bar is too brief compared to that of the star-forming process. Whatever the reason, the bar in M83 is not unique. Most bars in spiral galaxies are made up of old stars, with very little star formation except near their ends. This observation is worth keeping in mind when one is evaluating the UV-emitting structures of distant galaxies at high redshift (see Chapter 16).

M94, the Nearest Ringed Spiral

Except for the Milky Way itself, M94 (NGC 4736) is the closest galaxy to sport a prominent inner ring of hot blue stars and associated nebulosity (see Plate 5). Located about 15 million light-years away in the constellation Canes Venatici, this Sab-type galaxy features a visibly dominant bulge and inner disk of old red stars, with tightly wound blue spiral arms in the outer disk (see Plate 1).

The inner starbursting ring is paired with a much fainter outer ring of older stars. Linking and perhaps generating these two annular features is an oval distortion in the underlying stellar disk (see Plate 5). The fact that such ring-bar pairings are fairly common hints at some underlying dynamical connection. The most successful model to date has bar-mediated resonances in the disk orchestrating the location of the inner and outer rings. These resonances also help to explain other features, including the UV-bright knots seen on opposite sides of M94's nucleus, just beyond the ring (see Plate 5).

The edge of the oval disk appears to coincide with the so-called co-rotation radius, where the stars and gas rotate in step with the density wave. Inside of co-rotation the material overtakes the density wavefronts, while outside of co-rotation the wavefront outpaces the revolving constituents. M94's faint outer ring and starbursting inner ring would then correspond to resonant radii where the material radially oscillates once for every encounter with a density wavefront. Even the UV-bright knots at intermediate radius may have resonant pedigrees, tracing where the stars and gas oscillate twice for every passage through a density-wave crest.

Similar bar-mediated resonances are thought to explain most of the ringed-barred galaxies that are observed today (see Plate 20). Annular resonances may also have played an important historical role in building up the inner disks and central bulges of giant spirals, although their precise role in this evolutionary process has yet to be worked out. When present, circumnuclear resonant rings can impede the inward flow of material that would otherwise feed the nuclei, and these sorts of roadblocks could explain the relative rarity of intense nuclear activity in giant spirals.

In the case of M94, all of these evolutionary scenarios seem plausible. The rate of star formation in the UV-bright ring, when integrated over a 10-billion-year period, is sufficient to have built up all of the stellar mass that is observed in the inner disk and bulge. Meanwhile, nuclear activity in this centrally concentrated giant spiral is barely detectable, despite the copious amounts of gas that pervade the inner disk.

Centaurus A, the Nearest Giant Elliptical

Leaving the giant spirals behind, we now find ourselves in very strange territory. Like their dwarf siblings, the giant ellipticals don't even pretend to show ordered circular rotation in a common plane. Instead, the stars in giant ellipticals swarm in highly elongated elliptical orbits without strong preference for any particular plane. Our diagnostic challenge is made even worse by the propensity of each giant elliptical to show pathological tendencies. So it is with the nearest elliptical colossus, Centaurus A (see Plate 21).

Situated only 11 million light-years away, Centaurus A (NGC 5128) received its name by being the brightest radio source in the southern constellation of Centaurus. In the *Revised Shapley-Ames Catalog of Bright Galaxies*, it is classified as an S0+S peculiar galaxy, yet in the *Third Revised Catalogue of*

Galaxies it is given an E2 peculiar classification. Already, we are in trouble; a quick glance at Plate 21 and Figure 1.1 shows why. Besides having an overall elliptical shape, Centaurus A sports a dark girdle of obscuring dust. This silty raiment is thought to be the debris of a spiral galaxy that had the misfortune of merging with the larger spheroidal system. At infrared wavelengths, the innermost part of the captured galaxy shines through the outer dust belts, revealing a two-armed barred spiral (see Figure 1.1).

Compared with other, more distant, giant ellipticals, Cen A seems to represent a special moment in the lives of these systems. Most of the other giant ellipticals are not caught in the act of devouring their neighbors. Yet they probably have all been involved in galactic cannibalism at some point(s) in their lives. The telltale signs are legion, including multiple nuclei, boxy cores, and alternating shells of stars in the outer extremities of these giant galaxies. These shells are thought to be left behind by the merging galaxy as it spirals into the gravitationally dominant galaxy. More recent observations have revealed small dusty disks inhabiting many giant ellipticals—perhaps generated in prior feeding episodes. In Chapter 10, we will look at the earlier stages of the merging process, where the two interacting galaxies are still distinct systems. The optical luminosity of Centaurus A is truly astounding. With an absolute blue (*B*-band) magnitude of $M_B = -21.7$ mags, it is as bright as the brightest face-on spirals, and would be considerably brighter were the obscuring disk not there. Centaurus A is also noted as being the nearest active galaxy. Although its nucleus is hidden from view, giant lobes of radio emitting plasma can be traced back to the obscured nucleus (see Plate 21). The outermost lobes extend outward for 800,000 light-years, spanning a whopping 10 degrees on the sky (see the all-sky radio map in Plate 7). In Chapter 11, we will take a closer look at this energetic activity and its connection with the nuclear engine that drives it.

10

INTERACTING AND STARBURSTING GALAXIES

In the previous chapters, we have considered galaxies as individuals, much like the "island universes" first proposed by Immanuel Kant some 250 years ago. Yet without exception, we have found that these immense systems are invariably under the influence of neighboring galaxies. From the satellite dwarfs buzzing about the Milky Way to the hapless spiral being digested by Centaurus A, the nearest galaxies manifest a myriad of mild and not-so-mild interactions. Indeed, the sundry galaxy interactions that have occurred over cosmic time are in many ways responsible for having generated the types of galaxies that we see today.

Some interactions are so obvious that they can be spotted by just looking at pictures of closely paired galaxies. Fritz Zwicky was one of the first astronomers to make note of these odd pairings. His early drawings from the 1950s show some of the characteristic features endemic to strong interactions: tidal "tails" leading away from the gravitating pairs, with faint "bridges" often linking the interlopers (see Figure 10.1). Like dancers with their left arms joined and right arms in flamboyant opposition, the galaxies twirl about one another in a fateful fandango, culminating in the closest of embraces.

In 1966, Halton Arp employed the 5-meter (200-inch) telescope at Mount Palomar to create a comprehensive photographic atlas of peculiar and interacting galaxies. Consequently, some of the most famous interacting systems are known by their Arp numbers (see Figure 10.1). Let us begin with

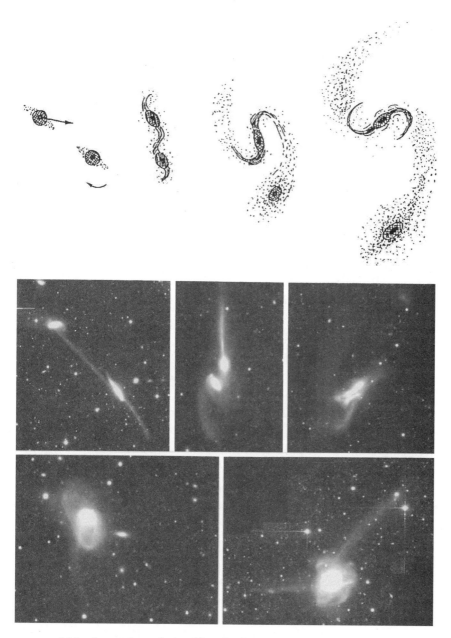

FIGURE 10.1   Interacting galaxies. Top: An interaction sequence as envisioned and sketched by Fritz Zwicky, based on photographs that he took on Mount Palomar, circa 1953. Middle and bottom: A merger sequence assembled from "snapshots" of galaxies taken in various stages of interaction. The galaxies are (middle row, from left to right) Arp 295, the Mice (NGC 4676, Arp 242), and NGC 520 (Arp 157); and (bottom row, left to right) NGC 3921, and Atoms for Peace (NGC 7252, Arp 226).

one of the more subtle and exquisite cases, the lopsided spiral M101 (Arp 26), and then examine increasingly deviant examples of the interacting genre, the Whirlpool (Arp 85), the Antennae (Arp 244), and the Cartwheel (a galaxy listed in the *Catalogue of Southern Peculiar Galaxies and Associations* by Halton Arp and Barry Madore as Arp-Madore 0035–335).

One frequent outcome of strong interactions is the triggering of spectacular starburst activity. To better understand this powerful and transformative phenomenon, we will end the chapter by scrutinizing two nearby starbursting galaxies. The nearest global starburst, NGC 1569 (Arp 210), appears to be going through its dramatic paces without the aid of any major interactions. The nearest nuclear starburst, M82 (Arp 337), exemplifies the more common circumstance, since it is interacting with both a giant spiral and a smaller amorphous galaxy.

Arp's compilation of peculiar and interacting galaxies includes some of the most visually dramatic systems in the local Universe. Although sophisticated computer simulations are necessary to delineate the evolution of these galaxian mixups, the computer models are all based on the same physics that drives the tides on Earth: Newtonian gravity.

M101 (NGC 5457; Arp 26), the Cockeyed Pinwheel in Ursa Major

One of the largest galaxies in the nearby Universe, M101 is a sprawling Sc-type spiral whose disk spans the equivalent of two Milky Ways. Arp listed it as number 26 in his *Atlas of Peculiar Galaxies* by virtue of its asymmetric disk and odd-shaped spiral arms (see Figure 10.2). In the early 1990s, high-resolution images of M101 revealed several Cepheid variable stars. Analysis of the Cepheids' periods and apparent magnitudes yielded an estimated distance of about 22 million light-years. At this distance, each of the supergiant H II regions in the arms of M101 is as big and luminous as a starbursting irregular galaxy unto itself.

Ultraviolet images of M101 have shown that its arms do not trace logarithmic spirals (where the polar angle scales with the logarithm of the radius), but instead consist of straight-arm segments that abut one another—often at angles of 120 degrees or so (see Figure 10.2). These crooked arms are seen in simulations where massive perturbers within the disk generate density wavelets, whose resonant interactions with the orbiting gas can produce the straight-arm structures. The supergiant H II regions in the disk of M101 may provide the perturbations necessary to drive this sort of dynamic. But what brought on the supergiant H II regions?

One clue can be found in the most luminous of M101's H II regions,

FIGURE 10.2 Hints of perturbation in M101. Top: As viewed at blue wavelengths, M101 looks like a giant grand-design Sc-type spiral galaxy. The luminous knot to the left (east) of the left-most spiral arm is the supergiant H II region NGC 5471. Dwarf companion galaxies lie farther out, beyond the field of view. Bottom: At ultraviolet wavelengths, the inner disk and bulge disappear, while the arms appear disjointed rather than smoothly spiraling.

NGC 5471. Deep UV imaging has shown that this "hypergiant" H II region is linked to the rest of M101 by a faint spiral arm whose curly tail wraps around NGC 5471. The faint arm and massive H II region at its tip were most likely spawned by some tidal interaction with an external object. Likely instigators are the two satellite dwarf galaxies—NGC 5474 and NGC 5477—whose masses and proximity to M101 qualify them as significant perturbers. Simulations involving a close passage of the more massive dwarf NGC 5474 about 500 million years ago successfully replicate M101's lopsided disk and outermost spiral structure. Other simulations predict massive condensations growing at the ends of tidal tails, consequences that are similar to the curly tail and hypergiant H II region in M101.

So it seems that both internal and external interactions are necessary to account for the odd features in M101. Several other Sc-type galaxies show crooked arm morphologies and disk asymmetries, which suggests similar interactions at work. These sorts of tidal dynamics are as subtle as they are pervasive—competing and cooperating with whatever density waves and stochastic processes may be active at the time. As we will see, stronger tidal interactions lead to more drastic alterations of the galaxies involved.

M51 (NGC 5194 + 5195; Arp 85), the Whirlpool

Located near the handle of the Big Dipper, M51 is one of the most recognizable galaxies (see Plate 22). Its robust spiral structure and prominent companion are emblematic of an interacting dynamo in progress. As far back as 1850, William Parsons, the third Earl of Rosse, was observing these outstanding features with his giant reflecting telescope. Nowadays, the big challenge is to ascertain the dynamical history between the Sc-type spiral NGC 5194 and the interloping amorphous galaxy NGC 5195.

The first major advance came in the 1970s, when Alar and Juri Toomre of MIT created a computer model of the interaction. Using two point masses surrounded by a few hundred test masses to simulate two interacting disk galaxies, they were able to make a crude movie showing the close encounter and subsequent generation of spiral arms along with tidal tails and bridges. They also predicted that the two galaxies would ultimately merge into something resembling an elliptical galaxy.

Since then, far more sophisticated computer simulations have confirmed the Toomre brothers' seminal work, while bringing up new issues. One particular concern is the effect of the interaction on the innermost spiral structure: How can the external perturbation work its way into the inner disk? Perhaps NGC 5195 orbited close to NGC 5194 twice in the past few billion years. The first encounter could have tidally modified the disk of NGC 5194, while the second flyby could have excited density waves throughout the altered disk.

Density-wave dynamics in NGC 5194 are clearly evident in offset distributions of dust lanes, Hα emission from the H II regions, and blue starlight along the spiral arms (see Plate 22). Displacements in these various tracers of recent star formation indicate a spatial progression between the star-forming clouds of gas and dust, the hottest O-type stars that ionize the hydrogen gas, and the slightly older and cooler B-type stars that dominate the observed blue light. A propagating density wave would set up just this sort of spatio-temporal sequence across the spiral arms. So, despite the strong tidal alterations brought on by NGC 5195, the disk of NGC 5194 is still amenable to hosting spiral density waves. Indeed, NGC 5195 may be doing double duty by exciting these waves.

The Antennae (NGC 4038 + 4039; Arp 244), a Merger in Progress

Among Arp's deviant galaxy pairings, the Antennae stand out as a magnificent example of what happens when two roughly comparable galaxies intertwine (see Plate 23). NGC 4038 and NGC 4039 have similar sizes and luminosities. Although it is difficult to tell today, they were probably both giant spirals just a few billion years ago.

What we see today are two giant commas, whose vast tails (or antennae) lead diametrically away from the grazing cores. At an estimated distance of 60 million light-years, the tidally distended system has a projected extent of 300,000 light-years—or the equivalent of three Milky Ways laid end to end. Measured radial velocities indicate that the southern tail is moving away from the southern core at about 100 kilometers per second. To attain such a huge size, the tail must have been ejected some 500 million years ago.

The Toomre brothers were once again the first to successfully simulate the encounter that led to this bizarre configuration. They began by numerically approximating two disk galaxies of equal mass, and then had the two galaxies approach on nearly colliding trajectories. Their computer model and more recent simulations that consider self-gravity and gas along with the stars indicate a future of ever closer encounters, ultimately leading to a complete merger.

Meanwhile, the encounter has led to a one-sided starburst, as can be seen in Plate 23. Because ultraviolet images of galaxies preferentially reveal the starbursting regions, those of the Antennae show only the northern half. If this galaxy pair were located in the distant Universe and thus at high redshift, its visible appearance would resemble the fetal shape that we see at zero redshift in far-UV images. We would have no idea that two galaxies were involved. Perhaps many of the peculiar-looking galaxies that we see at high redshift are also parts of interacting pairs like the Antennae. That is but one of many reasons why diagnosing the types of galaxies that inhabit the highly redshifted primeval Universe is a difficult challenge, fraught with uncertainty (see Chapter 16).

A closer look at the starbursting half reveals dozens of dense young clusters. From their extremely blue colors, they can be only a few million years old. And since their luminosities exceed that of 30 Doradus by factors of 2 to 20, they must have total masses of several hundred thousand Suns. Therefore, we may be witnessing globular clusters in the making. Somewhat older globular clusters have been found in another well-studied merging system—NGC 7252 (also called the Atoms for Peace galaxy, see Figure 10.1). The older ages are consistent with the more complete merging status of this system. Here, two disk galaxies probably began merging more than a billion years ago. We now see a chaotic flurry of shells, filaments, and tails in the stellar distribution, with counter-rotating components in the gaseous core.

An even more evolved outcome of the merging process can be seen in the giant elliptical galaxies Centaurus A and M87 (see Plates 21 and 27, respectively). Super-massive black holes reside in the centers of these two behemoths, the likely result of prior merging, or put less politely, cannibalistic activity. Fresh material is feeding these monsters, for we see them regurgitating powerful jets to great distances (see Chapter 11).

The Cartwheel (Arp-Madore 0035–335), Aftermath of a Direct Hit

Perhaps the strangest of galaxy interactions occurs when one galaxy plunges directly through the heart of another galaxy. Computer simulations of such an unlikely event predict expanding rings of gas and dust to form in the disk of the targeted galaxy—much like the circular ripples that form after one drops a rock into a pond. The expanding rings represent the disk's response to the hit-and-run galaxy. As the intruding galaxy approaches the disk, its gravity generates an inward-directed acceleration in the disk that intensifies and then weakens once the galaxy punches through and begins to recede. This perturbation acts much like a density wave, creating ring-shaped pile-ups of stars and gas in the disk.

Contrary to what happens in resonant-ring galaxies (see Chapter 9), the rings in these collisional systems are predicted to expand at considerable velocity. They are also more likely to be somewhat off center, because of the greater probability of slightly askew collisions. Finally, the collisional-ring galaxies should be closely accompanied by the colliding agents.

All of these conditions are present in the Cartwheel, the nearest and best observed collisional-ring galaxy (see Plate 24). At a distance of 400 million light-years, the Cartwheel can be considered a rarity in the local Universe. Its outermost ring has a diameter of 143,000 light-years, rotates at a velocity of 250 kilometers per second, and expands at 90 kilometers per second. The diameter and expansion velocity date the ring-producing collision to about 300 million years ago.

Studding the big ring are many supergiant H II regions, whose combined luminosities indicate starburst activity 50 times more intense than that in the entire Milky Way. Interior to the big ring are diffuse spiral "spokes" containing somewhat older stellar associations, along with gas shaped into rings or bubbles by recent supernova explosions. The spokes connect with an off-center inner ring, whose yellower color suggests even older stellar populations. This ring also shows dust lanes and comet-shaped structures that may trace gaseous shocks due to gas falling back at supersonic velocities. Finally, the offset heart of the Cartwheel contains a small ovoid core with a pointlike nucleus surrounded by bluish clusters.

Just beyond the Cartwheel are two smaller galaxies. Since both of them lie close to the minor axis of the Cartwheel, either one could have punched through its disk and so have generated the observed features. The H I in the Cartwheel extends out to an even more distant galaxy, however, which suggests that these two systems have a shared history. Meanwhile, the Cartwheel itself challenges us to predict what will become of it. Will its rings fade and its perturbed disk eventually relax into something resembling a normal spiral galaxy? Or will it continue to interact with its galactic provocateurs and so become even more peculiar? We are left musing on the pedigrees of the other strange and rare galaxian systems that inhabit Arp's *Atlas of Peculiar Galaxies*.

NGC 1569 (Arp 210), the Nearest Global Starburst

From the Orion Nebula to the Hubble Deep Field, intense star formation can be found energizing, transforming, and enriching myriad galaxian realms. Although this so-called starburst activity can be simply described in terms of massive stars and supernovae, the mix of radiative and mechanical processes that is powering the observed activity is far less well understood. Our ignorance becomes ever greater as we investigate the most powerful and distant starbursting systems (see Chapter 11). That is why the nearest starbursting galaxies are of special interest.

The dwarf irregular NGC 1569 (Arp 210) is the nearest example of a globally starbursting galaxy (see Plate 25). It is close enough to have its brightest stars resolved in groundbased images, and thus fairly accurate estimates of its distance can be calculated. The resulting figure of 7.2 million light-years places NGC 1569 just beyond the Local Group in the vicinity of the M81 group. Despite its obvious starbursting behavior, NGC 1569 has no close neighbors and hence cannot be part of some interacting system. Apparently, this galaxy is in the starbursting throes of its own doing.

Images of the red, Hα-emitting hydrogen gas vividly highlight the global nature of this starburst. In addition to discrete giant H II regions along its stellar bar, NGC 1569 sports an arm and two fainter arcs of fluorescing nebulosity. The excitation of this gas is probably caused by extreme ultraviolet emission from O-type stars in the bar. The generation of the nebular arm and arcs, however, is most likely the result of multiple supernova blasts having occurred. Considerations of likely outflow geometries and kinematics lead to an estimated age of 10–30 million years for the eruptions, with a total energy requirement equivalent to thousands of supernovae. Vestiges of this explosive activity include gaseous filaments near the stellar bar with radial velocities of hundreds to thousands of kilometers per second, along with diffuse X-ray emission from gas that has been shock-heated to millions of degrees (see Plate 25).

Closeups of the stellar bar in NGC 1569 reveal two dense blue superclusters. From the spectra of these clusters, one can infer that their light is dominated by short-lived supergiant stars of the A and B spectral types. The most massive O-type stars in the clusters probably died millions of years earlier—and in so doing, may have created the outflows evident in the gas. These clusters are five times more luminous than the 30 Doradus cluster in the Large Magellanic Cloud, and hence are prime candidates for young globular clusters.

Intense star formation continues today, as manifested by the young H II regions in the bar. But it is probably at a significantly lower level than that experienced during the galaxy's starbursting heyday—some 25 million years ago. By comparing the Hα fluxes from the ionized hydrogen gas in NGC 1569 with those from other galaxies, one can benchmark the intensity of the galaxy's current star-forming activity. NGC 1569 is forming stars at a rate per unit area that is 10 times greater than that observed in the Milky Way and the giant Sc spiral M101. But it is worth remembering that NGC 1569 is only a dwarf in comparison—one that happens to be overwhelmed with starburst activity. There are individual supergiant H II regions in M101 that are similar in size and luminosity to NGC 1569. If they were observed separate from their host galaxy, they would be dubbed starbursting galaxies in their own right.

One of the major lingering questions regarding NGC 1569 is the fate of its gas. Much of the gas that was enriched with metals during the starburst seems to be escaping from the galaxy. If this gaseous component is truly departing, never to return, then NGC 1569 will remain a metal-poor dwarf with meager prospects of evolving chemically. If, however, the outflowing enriched gas eventually falls back into the galaxy, then NGC 1569 will enjoy a new lease on life. Indeed, cycles of starbursts followed by fallow periods and renewed starburst activity may hold the key to understanding the natural histories of dwarf irregular galaxies such as NGC 1569. This sort of cyclic behavior may also have its counterparts in galaxies observed at high redshifts and hence at great lookback times, when globally starbursting behavior appears to have been more common.

M82 (NGC 3034; Arp 337), the Nearest Nuclear Starburst

The grand-design spiral M81 may appear tranquil (see Plate 18). But its exquisite symmetry belies its participation in a closely interacting group of galaxies. Radio mappings reveal the pervasive H I gas that connects and feeds the various members (see Figure 10.3). In addition to M81, the system includes the amorphous galaxy NGC 3077, the dwarf Holmberg IX, and the starbursting amorphous galaxy M82. Despite the rich gas content and tidal turmoil, only M82 is starbursting now.

When viewed through an 8-inch amateur telescope, M82 looks like a fuzzy cigar, providing a palpable contrast to the centrally concentrated and disky appearance of M81. The kinematics of the stars and gas in the "cigar" show a regular pattern that can be easily explained if M82 is a small late-type disk galaxy, seen nearly edge-on. So far, there is little to suggest anything special. Observations at other wavelengths tell an entirely different story, however.

Far-infrared measurements of M82 rack up 30 billion Suns worth of luminosity. Stellar heating of the thick dust seems to account for what is observed, implicating a massive cache of newborn hot stars. Near-infrared imaging reveals the starlight itself, from what appears to be a nuclear or circumnuclear starburst deep inside the dust-roiled disk. Radio observations show strong and extended synchrotron emission, whose origins can be explained by multiple supernova detonations releasing high-energy electrons into the outblowing interstellar medium.

Deep imaging in the light of ionized hydrogen, nitrogen, and sulfur re-

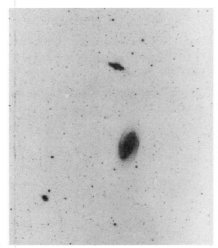

FIGURE 10.3   Visible view of the starlight (left) and radio view of the H I gas (right) in the interacting M81 group of galaxies. From bottom to top, the galaxies and their respective types are NGC 3077 (amorphous), M81 (Sb), and M82 (amorphous). The radio view reveals the H I gas that pervades and connects the interacting galaxies. Only M82 is starbursting now (see Plate 26).

veals the starburst in all its glory (see Plate 26). Here we see vast filaments of ionized gas streaming away from the dust-curdled stellar body of the galaxy. Doppler shifts in the line-emitting gas indicate outflow velocities of about 500 kilometers per second. At this speed, the observed filaments are most likely tracing gas that was blown out of the galaxy some 5 million years ago. The outflowing gas in the filaments is exposed to the enormous power of the central starburst, and so is being both photo-ionized by the hot stars and shock-excited by the supernova blast waves. The total energetics associated with this bipolar superwind is astounding, amounting to several million supernova explosions—all occurring within a region measuring less than 3,000 light-years across, and within a timespan of just a few million years. Constraining all this eruptive activity is a "dusty chimney": a thick donut of molecular gas and dust in which the most recent star formation is proceeding apace.

Compared to the star-forming intensity (rate per unit area) in normal spiral galaxies like the Milky Way, the star-forming intensity in M82 is about 400 times greater. The bulk of this starbirth activity can be explained by the much higher molecular gas densities that are present in the center of M82. But they do not explain everything. A measure of the star-forming efficiency can be obtained by dividing the star-formation rate by the available mass of fueling gas. In normal galaxies, the star-formation rate pretty much tracks with the available gas—yielding a nearly constant efficiency of about 50 percent conversion of gas to stars every billion years or so. Taking the reciprocal of the efficiency gives an estimate of the characteristic timescale for significant gas consumption, which in normal spirals amounts to a few billion years. By contrast, the star-forming efficiency in M82 is 16 times higher, and the corresponding gas depletion timescale 16 times shorter. Clearly, we are dealing with bursting conditions in M82.

These calorimetric considerations of M82's prodigious starbirthing lead us to question how the efficiency can be so high. Perhaps other agents besides the ambient gas density are governing the rate of stellar production. Viable candidates for triggering or catalyzing the star formation include cloud-cloud collisions in the gas-rich core, pressurization of the cold molecular clouds by the hot gas produced in prior supernova explosions, and shock compression by the supernova blasts themselves.

One fact we cannot elude—an incredible amount of gas had to stream into the nucleus in a very short period of time to make what we see today. Perhaps a central bar in M82 has helped to orchestrate this deluge, or perhaps prior close interactions with M81 and NGC 3077 prompted the ambient gas to plummet into the heart of M82. Similar questions face us when we try to understand the even more powerful ultra-luminous infrared galaxies and active galactic nuclei that are seen at much greater distances.

11

THE MOST
POWERFUL
GALAXIES

Among the giant spiral, elliptical, and interacting galaxies there are a few sources whose power outputs are so great that they beg explanation. Some appear to be overgrown versions of the starbursting galaxies that are observed nearby. The most powerful of these are known as ultra-luminous infrared galaxies (ULIRGs), by virtue of their intense mid- and far-infrared emissions. Others seem to require completely different powering mechanisms. These galaxies are noted for having brilliant nuclei, sometimes with fantastic jets emanating from them. Radio galaxies, Seyfert galaxies, blazars, and quasars are just a few examples of this genre. Despite their remarkable variety, these galaxies are lumped together as having active galactic nuclei (AGN). A common mechanism may in fact be powering these sundry systems—one that has a super-massive black hole and surrounding accretion disk as its engine. Moreover, an evolutionary progression from the starbursting ULIRGs to the blazing AGN may help to explain the coexistence of these phenomena at early epochs. In the following sections, we will pay our respects to the diverse embodiments of the most powerful elite—beginning with the dusty ULIRGs, moving on to the vast radio-lobe galaxies, and culminating with the brilliant quasars.

Ultra-Luminous Infrared Galaxies

When the Infrared Astronomy Satellite (IRAS) was launched in 1983, its primary aim was to map the entire infrared sky. One of the first data products to result from this effort was a catalogue of pointlike infrared sources. Although most of the sources turned out to be low-temperature stars and dense star-forming clouds in our own Galaxy, there remained a surprisingly large number of extragalactic sources. The search for optical counterparts ultimately yielded a vast harvest of galaxies, some of which had not been recorded previously. Among these galaxies were many star-forming systems like our own Milky Way. But there remained a few systems whose emission was almost entirely in the infrared part of the spectrum. Follow-up spectroscopy of these infrared-bright objects revealed redshifts that placed them far from us—with infrared luminosities greater than a trillion Suns, or the equivalent of a hundred Milky Ways.

Features in the infrared and optical spectra of most ULIRGs indicate copious star formation in progress. The great majority of the stellar radiation from this starburst activity is absorbed by the thick dust and reradiated at infrared wavelengths. Starbirth rates of 300 Suns per year are not unusual, outpacing the birthrate in the local starbursting galaxy, M82, by factors of 30 or more. Millimeter-wave observations of these systems reveal dense conglomerations of molecular gas. Despite the massive pile-ups of gas, the luminosities of these galaxies are still far greater than can be produced by "standard" starburst activity. Some of the excess luminosity might be triggered by galaxy-galaxy collisions. The rest might be generated by underlying AGN activity that warms the dust in the dense galaxian cores.

Recent imaging with the Hubble Space Telescope and groundbased observatories indicates that ULIRGs are often parts of strongly interacting systems (see Figure 11.1). Such close encounters were more likely when the Universe was in a denser state, a fact that may explain the greater numbers of ULIRGs that have been found at large lookback times. Indeed, ULIRGs significantly outnumber quasars and all other sources having luminosities greater than a trillion Suns. The dust-enshrouded ULIRG phase appears to have lasted longer than the more exposed phases of intense starburst and AGN activity—perhaps because it came first. Deep far-infrared observations will be necessary to resolve this question of precedence between ULIRGs and AGNs.

Radio Galaxies

The field of radio astronomy began in the 1930s with Carl Jansky's crude antenna at Bell Laboratories in New Jersey, a clumsy contraption made of wood and pipe. By the 1940s, the Sun, Jupiter, and the Milky Way were shown to be discrete sources of radio emission. Technological advances driven by World War II led to more elegant radio dishes and more sensitive

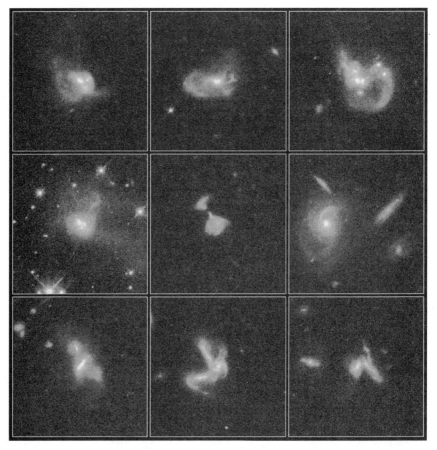

FIGURE 11.1 Ultra-luminous infrared galaxies (ULIRGs) are often associated with multiple systems of colliding galaxies. These images were obtained with the Hubble Space Telescope as part of a 3-year survey of 123 ULIRGs within 3 billion light-years of Earth.

receivers, which in the employ of radio astronomers revealed other sources in the sky. These were first thought to be radio "stars." The realization that these objects were distant and powerful galaxies came much later.

In the early days of radio astronomy it was not easy to figure out what was emitting the radiation that was being detected. Because radio waves are very long, typically about the length of this book from peak to peak, images received by radio telescopes are very blurred. The resolution of any kind of telescope depends on the ratio of the wavelength to the telescope aperture; so to improve the resolution at a given wavelength, one must increase the telescope size. But even giant radio telescopes that are much larger than the largest optical telescopes typically have poorer resolution than even the smallest of optical telescopes. Therefore, in the 1950s, when radio telescopes were still relatively small single dishes, it was difficult to see sources clearly and to be sure of their exact positions. Radio astronomers knew roughly where their objects were; they could at least say in which constellation they resided. But without more accurate positions it was difficult to figure out just which objects were giving out such surprisingly strong radio signals.

The First Galaxian Radio Sources

Given the tremendous uncertainties in position, the first catalogued radio sources were named after their constellations. The brightest was assigned the letter A, the next brightest B, and so on. Thus, the brightest radio object in the sky was named Cassiopeia A, or Cas A for short. Other early discoveries were Cygnus A (Cyg A) and Centaurus A (Cen A).

The first indication that some of these radio sources might be distant galaxies came in 1949, when Australian astronomers identified the strong radio source known as Centaurus A with the peculiar galaxy NGC 5128 (see Plate 21). The question was settled in 1954, when Palomar Observatory astronomers used the 200-inch telescope to search for visible objects that might coincide with other bright radio sources. After considerable effort, they found several peculiar objects that were good candidates and tried to figure out what they might be. One of them, a faint, strangely shaped clump at the position of Cygnus A, turned out to have a spectrum like that of a distant galaxy, whose redshift indicated a distance of about 700 million light-years. Other candidates were also galaxies with peculiar structures of one sort or another. Astronomers concluded that galaxies can be strong radio sources—but only very strange galaxies.

Galaxies in Collision?

Some of the first radio sources looked like two galaxies that might be colliding, and it seemed that galactic collisions might be the violent cause of the tremendously energetic events implied by the radio noise. But other radio galaxies showed no signs of collision. And besides, it was thought that simple encounters between galaxies would not produce the sorts of violent reactions that could give rise to strong radio emission.

The stars in a galaxy are so far apart compared with their diameters that a galaxy is nearly empty, as far as stars are concerned. The collision of two galaxies is not likely to involve a single collision of two stars. The galaxies will merely merge and then reemerge as they pass through each other. At first glance, therefore, it seems unlikely that a collision of galaxies would be a very interesting event, certainly not a spectacular one that would generate a strong radio signal.

But three possible kinds of events can occur that might make a big difference. First, a collision can produce violent tides that tear the galaxies apart; second, the collision of gas and dust in the two galaxies can lead to a hot, shocked medium that can radiate its alarm; and, third, in the case of galaxies of quite different masses, the larger one can swallow the smaller. For each of these kinds of events there are suspected candidates—some described below. But in the years immediately following the apparent discovery of colliding galaxies, the whole idea died a sudden and, as it turned out, undeserved death.

As more radio galaxies became identified, some of them turned out to be so bright that there was no obvious way that enough energy could be generated by a collision. Others consisted of a single galaxy, and often the solitary object was a nearly gasless elliptical. As it became clear that something besides a collision had to be thought up to explain most radio sources, other hypotheses began to flow from the pens of theorists. Among them were models that involved antimatter galaxies, magnetic flares, accretion of intergalactic matter, formation of new galaxies, chain-reaction supernovae, formation of new matter, and the action of a central, super-massive object. In the years since, this list has been pared down. New telescopes, new techniques, and thousands of new radio galaxies gradually disqualified almost all of the hypotheses. At present, though astronomers still do not know exactly what causes radio galaxies, the choices have been narrowed down to just about one possibility, the only one that seems capable of generating the immense amounts of energy and causing the many weird shapes and strange features that characterize radio galaxies.

In this model, a massive object, probably a black hole, resides at the galaxy's center, causing havoc as it inexorably sucks in the matter around it, matter whose final scream we hear in the radio spectrum. In order not to prejudge the case, however, astronomers do not usually speak of this thing as a black hole, but use some less specific term, such as a super-massive object. Some astronomers, eager to preserve their neutrality while recognizing the remarkable nature of this object, call it the central monster.

The Monster at the Center

Interferometric arrays of radio telescopes show that there is a tiny, central, powerful object in most active galaxies (those that emit radio or X-ray radiation or both). Intercontinental interferometers have demonstrated that these central objects must be very small indeed, less than about 1/10 of a light-year across. They seem to be the source of great jets of material and radiation, usually emitted in two opposing directions. In some galaxies these jets extend outward from the nucleus in a compact double formation, detectable at optical, radio, and X-ray wavelengths (see Figure 1.1). In others, the jets extend far beyond the visible galaxy, reaching out into the intergalactic space (see Figure 11.2).

The radiation making up the jets is characteristic of the kind emitted when electrons are moving at near-to-light speeds through a strong magnetic field. This light is called synchrotron radiation because it is seen by physicists as a glow in high-energy particle accelerators, known as synchrotrons. When the light of radio galaxies was first found to glow with these same characteristics, it was proposed that the radio radiation might be explained in this way, and now the evidence is overwhelmingly in favor of this idea. The jets that are seen in such galaxies as M87 (Virgo A) and NGC 5128 (Centaurus A) show up both in visible light and at radio wavelengths because they are the conduits of extremely high-energy particles (mostly electrons) that are hurtling out of the galaxies, dragging along strong magnetic fields as they go. The magnetic fields, in turn, constrain the outflows into the highly collimated jets that are observed (see Plates 21 and 27).

What is this thing in the center that can cause such spectacular eruptions? As mentioned above, the best explanation involves a super-massive object, probably a black hole. Black holes result from the complete collapse of a star or cloud that lacks enough rotation or internal energy to withstand the gravitational force of its own enormous mass. Einstein's theory of general relativity tells us that such an object will disappear from our view when it reaches a certain stage in its collapse, and from then on, no radiation can ever escape from it. Anything that falls into it will never be able to return. The density soon becomes incredibly large and strange relativistic effects commence, making little sense to our Earth-bound minds.

What we can detect is the immediate environment of the black hole, and this is very likely in the form of an accretion disk. As fresh material falls toward the black hole it rotationally and dissipatively flattens into a thin disk. Tremendous amounts of gravitational energy are converted into thermal motions in the disk, raising the temperature into the X-ray regime. Before this superheated material disappears down the hole, it has so much energy that it can squirt jets of relativistic electrons out from the rotational poles, and these produce the bipolar lobes observed in radio galaxies.

Super-massive black holes in the centers of radio galaxies are also implicated by the velocities of the stars and gas that are measured in the nuclei. In

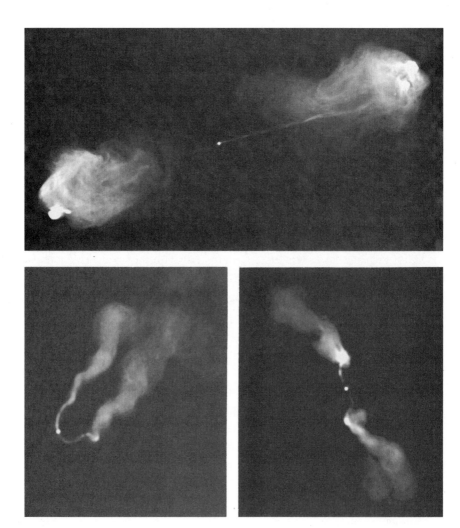

FIGURE 11.2   Radio galaxies, showing the extensive jets of radio-emitting plasma. These bipolar blowouts span 30,000 to 300,000 light-years in size, dwarfing their host galaxies by factors of ten or more. All images are based on observations with the Very Large Array (VLA) of radio telescopes. Top: A radio image of Cygnus A shows highly collimated jets piling up at their extremities, where they are impacting the intergalactic medium. Bottom left: The Head Tail galaxy, 3C129, manifests the effects of ram pressure by the ambient intergalactic medium. As the galaxy plows into the tenuous medium, its two jets are swept backward into parallel streamers. Bottom right: The giant radio galaxy Hydra A has oppositely directed plumes in an S-shaped configuration. This sort of sinuous structure implies coordinated changes in the orientation of the jets over time, perhaps due to a wobbling of the nuclear source.

the supergiant elliptical galaxy M87, for example, orbital velocities of 750 kilometers per second have been measured within 60 light-years of the center (see Plate 27). These velocities indicate an object of 3 billion solar masses is inhabiting the nucleus and is most likely powering the observed radio and visible jet activity.

Food for the Monster

To account for the tremendous outlay of energy in a radio galaxy, some form of fuel must be found. Currently, infalling gas is regarded as the best candidate, since it can release sufficient gravitational energy to power the accretion disks and jets. But then where does this gas come from? Calculations indicate that at least one Sun's mass per year is necessary. What stars or other objects are being sacrificed at this rate? In the ordinary course of events stars do not fall into the center of a galaxy no matter what is in there. They continue to revolve around it in their orbits unless acted upon by some extraneous force. In the early days of the galaxy, of course, there would have been some leftover gas in the disk that could fall to the center, not having the right amount of angular momentum to maintain an orbit. But this stuff should be all gone by now; instead, we must find a source that can be putting matter into the nucleus now, or at least very recently (to astronomers that means within the last billion or so years).

One source that looked reasonable at first was the matter lost regularly by stars as they evolve. We know that this process continually puts stellar material back into the interstellar gas so that it becomes enriched with heavier elements. Calculations of the rate of loss of stellar gas, however, indicate that only about 10 percent or less of the necessary quantity of gas could be acquired by the monster from this source. To satisfy its amazing appetite, we must look elsewhere for a provisioner.

A more promising possibility is galactic cannibalism. If a little galaxy should get too close to a big galaxy, the gas, dust, and stars in the disk of the smaller object would be tidally stripped away from its nucleus early on in the drama and would meld into the outer parts of the larger galaxy. The central nucleus of the smaller galaxy would not be easily disrupted, however, and should fall down to the center of the big galaxy. This source would easily provide the monster with enough full meals to keep up its activity for millions of years. Such collisions would not happen often, of course. But radio galaxies are fairly rare; only about 1 percent of the galaxies around us are markedly disturbed. Not too many close encounters would be needed to explain this rate. Moreover, radio galaxies are most common in galaxy clusters, where the chance of collision is greatest. We shall see that similar processes probably drive most other incarnations of the AGN phenomenon.

Seyfert and N Galaxies

As our knowledge of galaxies has progressed, we have seen many features common to a number of different kinds of unusual galaxies, some of which are not especially strong radio or infrared emitters. For example, in the 1940s Carl Seyfert discovered a class of galaxies that have extremely bright and broad emission lines in their nuclei. These are now called Seyfert galaxies and are the subject of many enlightening studies.

A nearby example is the Sb-type spiral galaxy NGC 1068, which is not only a Seyfert galaxy but also a radio source (see Plate 2). In its nucleus are very bright, hot gas clouds that show turbulent velocities of hundreds of kilometers per second. In energy, NGC 1068 is on the weak end of the range spanned by radio galaxies leading up to the most luminous quasars. NGC 1068 is the archetype of the Seyfert 2 category, where the spectral line emission is thought to arise from clouds orbiting the super-massive object in the nucleus at radial distances of about a light-year.

The Seyfert 1 category includes those AGNs with much faster motions, perhaps arising from clouds within a few light-days from the center. The closest example of a Seyfert 1 galaxy is NGC 4151, another early-type spiral. Whether a Seyfert 1 or Seyfert 2 galaxy is observed may depend on the inclination of the innermost disk to our line of sight. If highly inclined, a thick disk of gas and dust is likely to obscure our view of the innermost clouds, thus squelching any detections of the broad line emission that would denote a Seyfert 1 nucleus. The narrower lines from the slower-moving outer clouds would dominate our view, leading to a Seyfert 2 classification.

Another class of peculiar objects are the compact N galaxies. These AGN galaxies were first identified as a pervasive type among the more distant radio galaxies. We now recognize that many of their characteristics are quite similar to those of Seyfert galaxies; the N galaxies differ mostly by being farther away.

We have already mentioned another type of radio galaxy, the supergiant elliptical galaxy exemplified by Cen A and M87 (see Plates 21 and 27). These fat cannibals emit radio radiation from recent gorging on their cluster neighbors (see Chapter 9). They do not look like N galaxies; quite the contrary, they are bloated and diffuse. But many of the same mechanisms are at work in them, as well as in a variety of other galactic oddities. The champions, though, among a terrific line-up of characters, are the quasars.

Quasars

Radio and Seyfert galaxies are quite remarkable, but they hardly hold a candle to a group of enigmatic objects that were first detected in the 1950s. Radio astronomers, already familiar with radio galaxies as well as various kinds of nearby objects in the Milky Way, began to find something new—tiny, bright radio sources that were not identified with anything they had seen before. They appeared small, like stars, but were located in parts of the sky that were empty of any prominent galaxies, supernova remnants, or H II regions. As their positions were determined more and more accurately, the puzzle intensified. At first, these brilliant radio sources were seen to coincide with faint starlike objects, and so were identified as radio stars.

The First Quasi-Stellar Object

By 1960, radio astronomers were able to establish a precise position for one of these sources, called 3C 48. When they plotted this position on the Palomar Atlas of the sky, the only object at that position turned out to be a rather faint star with no obvious peculiarity.

The optical identification of 3C 48 with a star almost seemed to be a step backward. Years before, in the infancy of radio astronomy, most radio sources were thought to be stars and were called radio stars. As more knowledge accumulated, most of them turned out to be other kinds of objects—gas clouds, supernova remnants, galaxies. Suddenly, in 1960, there seemed to be a genuine radio star. But what kind of star could it be? What weird happening could make a star radiate such immense amounts of radio energy? To answer this question, the world's largest optical telescope was turned to look at it, first directly and then with a spectrograph.

The first deep optical photograph of the 3C 48 field, taken with the Palomar 200-inch telescope, yielded an exciting but also puzzling result. Right at the center was the "star," and as though to confirm that it was the real source, a faint wisp of light pointed to it like a finger. Clearly this must be the thing, but what was it? A spectrogram was obtained, but that did not help very much. In fact, the spectrum confounded everyone who looked at it. Instead of a continuous band of light of all different colors, like that of a star, the spectrum consisted of a faint band that was punctuated by a series of bright emission lines—all apparently in the wrong places. The chemical elements commonly found in star and gas clouds have characteristic patterns of emission-line wavelengths, and none of them seemed to correspond to the lines in the spectrum of 3C 48. The mysterious new radio sources were starlike in appearance, but they seemed to be made of unrecognizable material.

Initially, the discoverers called these objects quasi-stellar radio sources. This cumbersome name was all right when there were only two or three known, but as more and more of them were found, a shorter name was clearly needed. The decision had to be made between the initials, QSO, and a word made up from fragments of the words in the name, *qua*si-*stellar* *r*adio source, or quasar. Being catchy and exotic sounding, "quasar" soon became the popular choice, and it has since come to be used as a name for all kinds of nonastronomical things, from televisions to aircraft lights.

Redshifted Marvels

The next year saw the solution to the puzzling spectrum. In 1961 the Palomar telescope was used to obtain a spectrum of a 13th-magnitude quasar, 3C 273, the brightest-appearing of the quasi-stellar radio sources. When the photographic plate that recorded the spectrum was first developed, the resulting smear on the plate looked rather like the spectrum of 3C 48; there was a faint continuum superimposed upon a series of bright emission lines. But a pattern in the lines was recognizable that had not been apparent in the other spectrum. The lines were not in the expected place, but they were arranged in the right way, with the normal spacing and intensities. All, however, were greatly displaced toward the red side of the spectrum. These otherwise normal lines were redshifted so that they were almost missed. The same lines had, in fact, been there in the spectrum of 3C 48, too, but were redshifted even farther. In its case, the pattern was not evident because some of the familiar lines were redshifted right off the spectrum.

Redshifts of objects due to the Doppler effect were not unfamiliar to astronomers, of course. The astounding thing about the quasars was that the inferred recessional velocities were so immense. The largest velocities for stars in our Milky Way galaxy are about 400 kilometers per second (about 1 million miles per hour). The quasars, which looked like stars, had redshifts indicating recession velocities as large as 150,000 kilometers per second (about 325 million miles per hour). They could not be stars in our Galaxy, then, because they would be moving so fast that they would soon escape from the Milky Way to go hurtling out through intergalactic space. So what could they be?

Many hypotheses were suggested, some of them preposterous and some merely weird. Most were based on the assumption that the redshift was caused by the Doppler effect and that the quasars must, therefore, be moving rapidly away from us. Other ideas about the cause were also explored, including the possibility that a gravitational redshift might be involved. Albert Einstein had shown that in the presence of a very strong gravitational field the wavelength of light is lengthened in a way similar to the Doppler effect. This shifting is visible (just barely) in the spectrum of the Sun because of its large mass and is quite noticeable in the spectra of white dwarf stars, whose huge densities cause a significant gravitational redshift in the light emitted from their surfaces. But as time went on and quasars with larger and larger redshifts were discovered, the gravitational redshift model had to be abandoned, because no physical system could exist, for any length of time, with such an immense gravitational field as these redshifts implied.

In the decades since quasars were first discovered, huge amounts of data have been collected, but very slow progress has been made in our understanding. We now have identified over 3,000 quasars and find redshifts ranging from a few tenths to as large as 5 (these figures represent a comparison of the shift in wavelength to the rest wavelength, such that a redshift of 5 indicates a shift in wavelength that is five times greater than the wavelength as emitted by the source).

A Cosmological Explanation

The most likely explanation for the redshift has since become the so-called cosmological one: that quasars must be receding from us, like other galaxies, as part of the general expansion of the Universe, and that the enormous speeds indicated by the redshift would put them out among the most distant known galaxies. Some of the quasar redshifts were far greater than that of any galaxy yet measured—a fact that introduced yet another problem. How could these starlike objects be so remote?

In hindsight, the reason that no galaxies were known with such large redshifts was simply that galaxies so far away were too faint to be seen and measured. This implied that the quasars must be much brighter intrinsically than even the brightest galaxies. The luminosity of 3C 273, for instance, is calculated to be about 100 times that of a normal giant galaxy. Now we do see galaxies out among the quasars, and they are indeed faint by comparison.

Another conundrum was the discovery that many of the quasars were variable sources. For example, it was found that 3C 273, which had been inadvertently recorded on sky-survey plates at Harvard Observatory for over 50 years, had been varying in brightness irregularly over that interval of time. Subsequent monitoring of other quasars showed them to vary also, some by factors of 10 to 100. In a few cases, the change was rapid, with variations happening in a matter of only a day (see Figure 11.3). This discovery introduced really serious problems into the cosmological interpretation of quasars.

An object that changes so rapidly cannot be very big. Light travels 1 light-day in one day, so if an object can show a large overall change in its brightness in so short a time, it must be smaller than 1 light-day; otherwise, any changes would be smeared out by the time it takes for the light from the far side to reach the near side (see Figure 11.3). A light-day is only about the size of the Solar System (Pluto's orbit is actually about half a light-day in diameter). How could an object only the size of the Solar System emit light 100 times more powerfully than a galaxy of hundreds of billions of stars? Nobody knew. Radio astronomers intensified the dilemma by using their newly developed intercontinental interferometric techniques to measure quasar diameters directly, also finding them to be very small and complex in structure.

The cosmological explanation seemed to be such an impossibility that many astronomers thought the quasars must somehow be local, after all. Perhaps they are stars shot out of our Galaxy in some way at enormous speeds. Or maybe the redshift is due to some new physical phenomenon other than the Doppler effect. This last possibility seemed to be supported when astronomers found apparent connections and alignments between two or more objects with wildly different redshifts. One suggestion was that these objects demonstrate some unexpected effect of aging on light, whereby the wavelength of light changes with time. But this is a maverick idea; at present most astronomers believe that the examples of alignments and connections are merely optical illusions, and that the objects are in fact at vastly different distances.

Quasar Hosts

The 1980s brought new vision to the study of the enigmatic quasars. Astronomical imaging equipment had advanced, so that the space immediately surrounding the brilliant center of a quasar could be perceived. The question had been asked before but could not be answered earlier: Is there a galaxy there, hidden by the brilliant, overexposed image of the quasar? Are the quasars at the centers of galaxies? Are they somehow related to galaxies with active nuclei, such as the Seyfert galaxies? If only we could see a faint galaxy image right up close to the quasar, surrounding it, we would know that the answer is yes.

Some of the first successful tests were carried out with the quasar 3C 273, one of the closest quasars and the one whose redshift was first obtained. Deep groundbased imaging showed a "fuzz" of light surrounding the quasar.

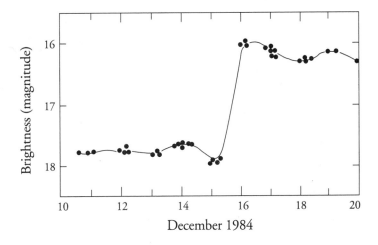

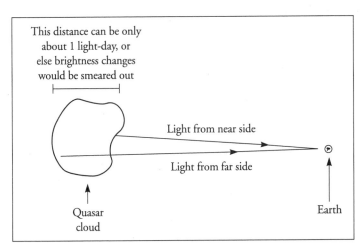

FIGURE 11.3   Quasar light fluctuations and their implications for the radiating source. Top: Variations in quasar luminosities tend to occur rapidly—sometimes within less than a day. Bottom: The observed timescale of brightness fluctuations can be used to limit the size of the emitting source. In this case, the quasar must be smaller than a light-day, or else its brightness changes would appear smeared out by the different times it would take light to travel to us from its different regions.

Its surface brightness distribution resembled that of a normal elliptical galaxy. Moreover, spectra of this faint fuzz showed a redshift that was identical to the redshift of the quasar itself. Therefore, if the fuzz really is a galaxy, it is receding with a velocity that corresponds to its great distance, and therefore the quasar, too, must be at the cosmological distance and must be partaking in the cosmic expansion.

More recent imaging with the Hubble Space Telescope has revealed a

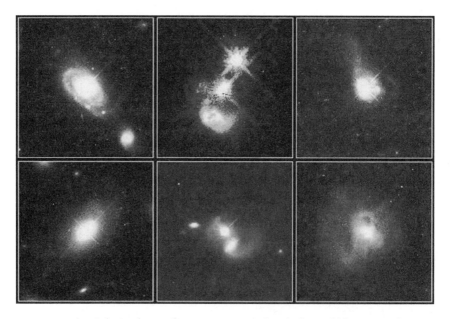

FIGURE 11.4   Galaxian hosts of quasars, as imaged with the Hubble Space Telescope. The hosts of some quasars appear as mixed up as the ULIRGs, and so may have similar collisionally induced pedigrees. Other quasar hosts look surprisingly normal.

strange menagerie of galaxy types associated with quasars (see Figure 11.4). Several appear to involve closely interacting galaxies—akin to the situation seen in the ULIRGs. Others appear remarkably normal, as if the galaxy doesn't know about the violent quasar activity in its heart. We finally know that quasars are terribly luminous, tiny, almost unbelievably energetic dynamos that are spouting off in the centers of some galaxies. The problem remaining is to find out why.

The Quasar Cloud

Even before they knew where quasars are, astronomers had built up a fairly detailed picture of the material that constitutes such objects. The spectra of quasars give enough information for us to conclude that there is a small, compact object at the very center, which is surrounded by a number of very hot clouds of gas, as well as some regions of cooler gas. Dust clouds seem to be interspersed throughout the nebular zone, and they have high velocities, as if they have been ejected from the more central regions. There are high-energy particles all around the central engine, moving in a strong magnetic field and emitting synchrotron radiation. It is this radiation that makes some quasars such strong radio sources. Although the strong radio-emitting sources initially led to the discovery of quasars, most quasars are radio-quiet. Consequentially, there must be many variations in the physical properties of these remarkable objects.

Some quasars do not show emission lines at all, as if they have no excited gas. With no lines in the spectrum, and hence no redshifts, it is not possible to learn the distance to these objects. These lineless quasars are called BL Lacertae objects, or blazars for short, after the prototype, which had been thought to be a variable star in our Galaxy. When BL Lacertae was detected as a strong radio source and when it turned out to display an optical continuum like that of a quasar, circumstantial evidence pointed to its identity as a quasar without emission lines. Now there are many other blazers known. We do not know the distances to many of them, except very roughly, based on their brightnesses at optical and radio wavelengths. For some very important BL Lacertae objects, however, distances are known by virtue of the redshifted spectral lines in their surrounding fuzz. They turn out to be among the distant galaxies and quasars, just as had been inferred from their other similarities to the quasars.

The Quasar Machine

One of the persistent difficulties posed by the quasars has been explaining the immense amounts of power they emit. Astronomers have tried many different things, including all the explanations put forward earlier to explain the radio galaxies. Only one mechanism seems to be capable of producing so much energy: gravitational collapse of a massive object, and the subsequent infall of material onto that object. Most astronomers believe that the centers of quasars, like the active centers of other galaxies, consist of massive collapsed objects, such as black holes. There is no real proof that a black hole is involved; it is merely the only thing anyone has thought of that could do the trick. If a quasar consists of a black hole, then its mass would have to be swallowing mass from the surrounding galaxy at a phenomenal rate of about 100 solar masses per year.

In the standard quasar model, the black hole is surrounded by a spinning accretion disk. The power comes from the release of gravitational potential energy, as hapless stars and gas clouds fall onto the disk and ultimately into the black hole. The disk supplies thermal energy to produce the light and probably involves magnetic flares, jets of lined-up particles, and other exotic phenomena. The high-energy gas particles form a plasma that can explain the X-rays emitted by some quasars.

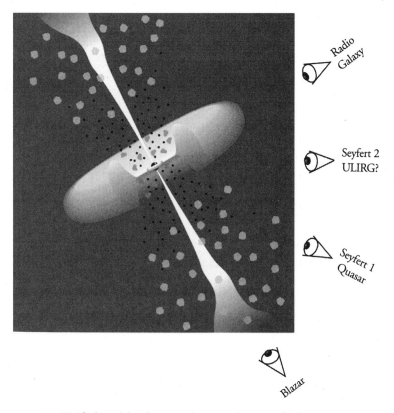

FIGURE 11.5 Unified models of active galactic nuclei typically feature a super-massive black hole, an enveloping accretion disk, a larger torus of gas and dust beyond the disk, irradiated clouds of hot and warm gas, and bipolar jets emanating from the accretion disk. The viewing angle and mass-accretion rate pretty much determine whether or not one detects a blazar, quasar, Seyfert 1, Seyfert 2, ULIRG, or radio galaxy.

If the central engine and accretion disk are further surrounded by a dense torus of dusty gas, then inclination effects play an important role (see Figure 11.5). Quasars seen pole-on would be seen as blazars; their light would be overwhelmed by the synchrotron-emitting plasma jet, so that no emission lines from the excited gas clouds would be detected. Quasars viewed a bit off-axis would personify the recognizable genre, with their innermost clouds in full view—radiating Doppler-broadened line emission. Quasars viewed farther off-axis would resemble the Seyfert 2 galaxies—with the broad-line regions obscured from view, but with the more extensive narrow-line regions showing above and below the dusty torus. Quasars viewed close to edge-on would display little nuclear activity but vast bipolar lobes of radio-emitting plasma.

This sort of unified model for quasars, blazars, Seyferts, and radio galaxies neatly encapsulates much of the variety associated with AGN galaxies. If such a model reflects reality, the inclination of the circumnuclear disk-torus system and the mass inflow rate pretty much determine what we see. To test such a simplifying paradigm, astronomers have been comparing the unified model's predictions with ever more detailed observations. So far—despite the tendency for each AGN galaxy to have its own special personality—the unified model continues to provide a useful context for understanding our particular view of these most powerful realms.

The Quasar Epoch

The redshifts of quasars bunch up at about 2 to 3, with hardly any quasars having redshifts greater than 5. There are also very few quasars with small redshifts, so that the distribution of redshifts is bunched at a recessional velocity that corresponds to an early epoch when the cosmos was roughly 25 percent of its present age. Quasars were much more common then than they are now and than they had ever been before. This peculiar fact has led some astronomers to name this interval, representing only about 10 percent of the history of the Universe, the quasar epoch. Why did quasars flare up back then? Why not earlier and why not now? These are some of the unanswered questions that still await some key insight or crucial observation, perhaps involving the ULIRGs, which also were more numerous during that epoch.

Some astronomers have suggested that the ULIRGs represent proto-quasars—with their starbursting cores still swaddled in obscuring dust. Inside the central cores of most ULIRGs, the gravitational collapse to a black hole is still in progress. Once that takes place, the resulting black hole could then begin to feed—with infalling gas clouds and hapless stars powering the quasar phenomenon. Some ULIRGs may be showing this sort of submerged quasar activity along with the more widespread starburst activity. Eventually, the violent repercussions of the blazing and erupting nuclear system would clear out the dust—laying bare a quasar that can be detected clear across the cosmos.

Alternatively, super-massive black holes could have formed first from the primordial chaos, with subsequent merger activity generating the quasar phenomenon. Further merging would have clothed the quasar in enough gas to spur rampant starburst activity. The resulting generation of metals from supernovae could then foster the rapid production of obscuring dust grains, thereby transforming the quasar into an ULIRG. Whether ULIRGs or quasars formed first remains a major topic of contention among scientists.

Quasars are the most energetic and spectacular examples of the main characters in this book, the galaxies. The fact that so much still remains unknown about them and that there is still so much controversy about their basic nature is an indication of the marvels yet to be unraveled as we continue to explore the Universe. The centuries have seen the expansion of our perceived horizon from the Earth to the stars to the galaxies. Now we stand at the edge of new and wondrous territory, the mysterious reaches of the quasars. And beyond them we can just glimpse the misty outlines of the primeval cosmos, before there were any galaxies at all. Next, we will ponder these incipient mists as we explore the overall nature of the Universe—how it is structured, how it expands, how it formed, and how it has evolved over cosmic time.

III

OUR
GALAXIAN
UNIVERSE

12

GAUGING THE GALAXIES

Over "mere" millions of kilometers, astronomers have established reliable methods for measuring distances with accuracies of better than one part in 10^6. But over the expanses of extragalactic space, where the distances are measured in terms of 10^{20} to 10^{24} kilometers, the methods are strained and the accuracy is low. Considering the immensity of the job, astronomers should feel fortunate that they are now able to argue about only 20 percent uncertainty. But the nature of science continues to drive astronomers to determine these distances much more accurately than that. Recent endeavors linking nearby Cepheid variable stars with more remote markers of distance hold great promise, so that astronomers may finally know the true scale of the galaxian Universe.

The Hubble Distance Scale

Starting in the 1920s with his pioneering work on galaxies in the Local Group, Edwin Hubble began an elaborate campaign to set up a distance scale that would extend to the edge of the observable Universe (see Figure 12.1). At first the scheme was crude, depending far too much on the assumption of uniformity. But it did lead to the first real appreciation for the vastness of the cosmos, and its method has been followed, in principle, by most astronomers in succeeding generations.

Hubble's first task was to measure distances to members of the Local

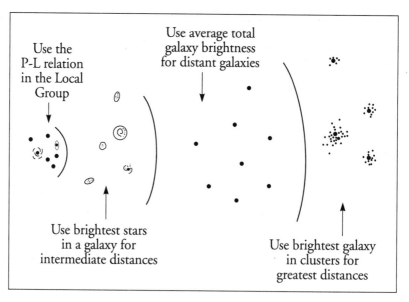

Use the
P-L relation
in the Local
Group

Use average total
galaxy brightness
for distant galaxies

Use brightest stars
in a galaxy for
intermediate distances

Use brightest galaxy
in clusters for
greatest distances

FIGURE 12.1 Hubble's scheme for measuring distances out to the faintest-appearing, most distant galaxies.

Group. He emphasized the galaxies M31, M33, and NGC 6822, having discovered Cepheid variables in all of them. His results for these three galaxies, combined with Harlow Shapley's data on the Magellanic Clouds and Walter Baade's perusal of other resolved galaxies in the Local Group, formed the baseline and the first step in Hubble's three-step cosmic distance scale.

We now have better measures of the distances to these galaxies, because of a revised period-luminosity (P-L) relation for Cepheids and improved brightness measures of faint stars. The resulting distances turn out to be two to three times farther out than Hubble first thought. But distances to the galaxies of the Local Group still form the framework upon which almost all distance scales are built.

Hubble's plan was, next, to use the nearby galaxies and their distances to calibrate luminosities of objects brighter than the Cepheid variables, so as to measure distances into farther reaches of space. After trying several kinds of objects, including red supergiants, star clusters, H II regions, novae, and so forth, he found that the brightest supergiant stars in a nearby galaxy all shared the same maximum luminosity, and that this maximum stellar luminosity was fairly uniform from galaxy to galaxy. Therefore, the apparent magnitude (or radiant flux) of a galaxy's brightest stars could be directly related to the galaxy's distance from the viewer.

A large collection of plates of many galaxies whose brightest stars were resolvable eventually provided Hubble with data that demonstrated to him the validity of this approach. He compiled measures of fluxes from the brightest stars in a long list of galaxies, and, as step 2, calibrated their distances by comparing these fluxes to those of the brightest stars in galaxies of the Local Group, whose distances had been previously determined from the Cepheid data.

Finally, Hubble's list of resolvable galaxies provided him with information on the intrinsic total luminosities of galaxies and their dispersion. Step 3, then, applied these luminosities to even more distant galaxies, beyond the range where individual stars could be resolved. This last step could be extended to the edge of the visible Universe, and it led to Hubble's most ambitious and grandest project: determining the size and the shape of the entire cosmos.

Redshifting Galaxies and the Hubble Relation

During the early 1920s, Hubble, V. M. Slipher, Milton Humason, and others had been taking spectra of galaxies and finding that some galaxies were moving with remarkable velocities, as indicated by the redward Doppler shifts of their spectral lines. In 1928, the cosmologist H. P. Robertson pointed out a correlation between the spectral redshifts, the inferred Doppler velocities, and the brightnesses of galaxies, whereby the recession velocities of galaxies are greater for the fainter objects. At about the same time, Hubble showed that this indicates an expanding Universe, with the velocity between galaxies being directly proportional to their distance.

Almost all galaxies showed redshifts, indicating velocities away from us; only a few galaxies in our Local Group and a few members of nearby clusters showed blueshifts. The Andromeda galaxy, for example, has a 300 kilometer per second velocity toward us, partly due to its membership in a gravitationally interacting group (perhaps it and we are in quasi-orbital motion) and partly due to our Sun's own orbital motion in the Milky Way. But the more distant galaxies are all receding. For example, among the galaxies of the cluster in the constellation Virgo the average velocity of recession is 1,000 kilometers per second.

The relationship between the radial velocities and distances of galaxies is referred to as the Hubble relation, which can be simply expressed as

$$v_r = H \times d,$$

where v_r is the velocity of recession as determined from the spectral redshift, d is the distance, and H is the constant of proportionality, also known as the

Hubble constant. Current workers usually refer to this constant as H_0, the subscript referring to the present value, since it may have had a different value in the past, depending on whether or not the expansion of the Universe has been constant or has accelerated or decelerated (see Chapter 14).

Sandage's Bigger Universe

A milestone along the path toward a reliable cosmic distance scale came in 1958, when Allan Sandage, Hubble's able protégé, gave a lecture on the occasion of his accepting the Warner Prize of the American Astronomical Society, awarded for outstanding achievement on the part of a young astronomer. The prize had resulted from his pioneering work in stellar evolution, but the lecture was on the distance scale and it caused something of a sensation. He showed some of the first results on this problem from the 200-inch Hale telescope atop Mount Palomar in California. Reworking Hubble's original sample of galaxies but using data from the giant Palomar telescope and more refined techniques, Sandage found several large errors in the early work, especially in the identification of a galaxy's brightest stars.

In 1936 Hubble had warned that identification of the brightest stars in a galaxy was not a trivial problem; star clusters, star clouds, and H II regions can all look stellar if viewed from far enough away. Sandage's startling conclusion in 1958 was that this problem had greatly affected Hubble's distances. Combined with earlier corrections, Sandage's new findings yielded a distance scale that was seven times larger than Hubble's 1936 scale. The Virgo cluster, for example, which Hubble measured to be 7 million light-years away, was found by Sandage to be 50 million light-years distant. The whole Universe was vastly larger than had been thought.

Sandage continued to pursue this problem with ever-increasing care, making each step more secure. In 1974 he and his colleagues began writing and publishing a series of papers that considered the problem once more, this time with several new intermediate steps that bolstered some of the weakest places in Hubble's bold plan. They were able to extend the use of Cepheid variables beyond the Local Group by training the 200-inch telescope on Cepheids in the spiral galaxy NGC 2403, one of a small group of galaxies 10 million light-years distant.

With two groups of well-calibrated galaxies (11 objects in all), they developed an auxiliary method that enabled them to address the awkward region that they called the twilight zone, between 50 and 300 million light-years. Here is where Hubble's work had gone wrong, where he had mistaken other kinds of objects for very bright stars. Sandage's approach used the apparent sizes of a galaxy's largest H II regions as the new indicators of distance. Because H II regions are resolvable out to the Virgo cluster and beyond, they provided an independent criterion, somewhat less subject to systematic identification problems than the brightest-star criterion. With the H II regions, astronomers could add a third group of nearby galaxies, centered on the giant spiral galaxy M101, to their calibrating base.

The Virgo Cluster Problem

Meanwhile, the Virgo cluster had fallen into disfavor as a source of information on the distance scale and the Hubble relation, for reasons that are quite remarkable. Although several astronomers had been arguing for years that we might be an outlying part of the Virgo cluster, convincing evidence did not appear until the middle 1970s. By then redshifts of thousands of galaxies had been measured throughout the sky—mostly as the result of a massive project carried out at the Harvard-Smithsonian Whipple Observatory in Arizona. The resulting spatial distribution of galaxy motions revealed a strong bias toward Virgo (see Figure 12.2).

Today, the preponderance of evidence indicates that our Galaxy and the Local Group are gravitationally attracted to a huge supercluster, of which the Virgo cluster is the densest and innermost part (see Chapter 13). Because our Local Group is gravitationally bound up with this supercluster, the measured velocities of the Virgo galaxies are complicated blends of the intra-cluster motions and the cosmic expansion velocity. This fact is itself an important clue to the question of how galaxy clusters form and how the mass of the Universe was organized into its present complexity. But for the purposes of determining cosmic distances, it is a complication.

A modeling of the recession velocities associated with unclustered galaxies and members of the Virgo cluster shows that the Local Group has a peculiar velocity toward the center of the Virgo supercluster of about 250 kilometers per second. Other combinations of distances and motions of galaxies have shown that the Virgo supercluster itself has a peculiar velocity of about 570 kilometers per second toward a more distant great attractor about 65 degrees away from Virgo. The cosmic background radiation, discovered in 1965 and identified as the relic radiation from the beginning of the Universe, also shows a variation in temperature with position, in roughly the same direction and of the same amount, which can also be attributed to these peculiar velocities. To discern the true scale and flow of the galaxian Universe, it is necessary to account for these sundry motions unique to our cosmic neighborhood—a task that continues to vex astronomers.

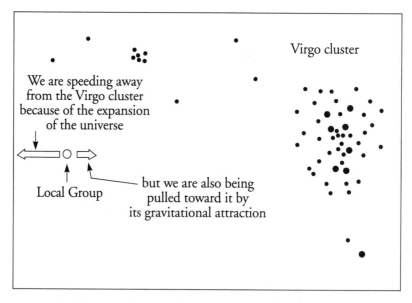

FIGURE 12.2   The Local Group of galaxies is gravitationally attracted to the Virgo cluster, and hence is receding from the cluster at a significantly lower velocity than expected from the overall cosmic expansion. A major problem for extragalactic astronomers is to disentangle the cosmological motions from the peculiar velocities caused by gravitating attractors such as the Virgo cluster.

Luminosity Effects

An historically important "rung" on the so-called extragalactic distance ladder is the luminosity of spiral galaxies, as inferred from structural clues (see Chapter 1). Building on his work with H II regions, Sandage and his colleagues used a morphologically based luminosity classification, similar to that first introduced by Sidney van den Bergh. After calibrating the luminosities in nearby groups of galaxies, Sandage and his colleagues determined the distances of 60 distant high-luminosity galaxies (of class Sc I) with radial velocities ranging from 3,000 to 15,500 kilometers per second. Comparing the distances with the recession velocities gave them an answer: the Hubble constant is even smaller (and thus, the Universe is larger) than previously thought. Hubble had derived a value of H_0 of 160 kilometers per second per million light-years (abbreviated as km/s/Mly), or 520 kilometers per second per megaparsec (abbreviated as km/s/Mpc, where 1 Mpc is a million parsecs, or 3.26 million light-years). Sandage had reduced that figure in 1958 to 23 km/s/Mly (75 km/s/Mpc). Now Sandage quoted a value of only 15 km/s/Mly (50 km/s/Mpc), with an estimated uncertainty of 10 percent.

Over the succeeding decades, the efforts to measure H_0 became ensnarled in controversy and generated some rather bitter feelings among the major players. Some scientists sided with Sandage's low value of H_0, while others adopted the values championed by Gerard de Vaucouleurs and his colleagues, which were up to twice as large. Many scientists chose to wait, reporting the results of their extragalactic research with allowance for the full range of possibilities. Through unstinting efforts by dedicated astronomers, this impasse was finally overcome in the 1990s.

Recent Advances

One of the most important developments of the last few decades has been the introduction of the H I linewidth method, using what is called the Tully-Fisher relation. This method involves carefully measuring a spiral galaxy's H I emission at the 21-centimeter wavelength. The width of the resulting spectral profile is a Doppler effect, caused by the overall dispersion of radial velocities in the rotating gas (see Figures 4.1 and 4.2). This spread in wavelengths and corresponding radial velocities turns out to be tightly correlated with the absolute (intrinsic) luminosity of a spiral galaxy—as quantified in the Tully-Fisher relation.

Although an empirical finding, the Tully-Fisher relation is explainable in terms of a galaxy's mass and the observable consequences of that mass. On average, the more massive a spiral galaxy is, the more rapidly it will rotate. It will also contain a greater number of stars and hence will shine more brightly. Once calibrated with nearby galaxies of known distance, the Tully-Fisher relation can be used to gauge galaxian distances of several hundred million light-years—well beyond the range of Cepheid variables and individual H II regions. The astronomer must first measure the galaxy's H I linewidth and apparent brightness. From the linewidth and the Tully-Fisher relation, the astronomer can surmise the galaxy's absolute luminosity. Comparing that luminosity with the galaxy's apparent brightness then yields the distance.

This far-reaching technique works best with infrared measures of luminosity, where problems with obscuration by dust are minimized. The resulting distance scale agrees rather well with Sandage's abandoned 1958 result, which said that the Hubble constant is about 23 km/s/million light-years (75 km/s/Mpc). A related attack on the distance scale uses detailed velocity curves, which give a full picture of how stars revolve in the spiral disk of a galaxy. The results of this analysis favor a value closer to about 70 km/s/Mpc.

Beginning with its launch in 1990, the Hubble Space Telescope has been

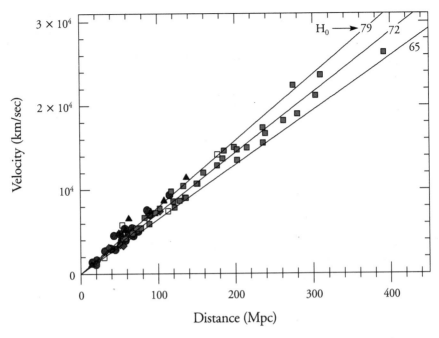

FIGURE 12.3   The Hubble relation between a galaxy's distance and recession velocity, as determined by team members of the Hubble Space Telescope Key Project to measure the Hubble constant, H_0. Their resulting value for H_0 is 72 km/s/Mpc with an estimated uncertainty of 8 km/s/Mpc in either direction. The five different kinds of data points denote different techniques for measuring the distances to galaxies. Measurements of Type Ia supernovae (filled black squares) provide the most far-reaching distances.

dedicated to solving the cosmic distance scale problem. As one of the HST's so-called key projects, the effort to determine the Hubble constant to better than 10 percent accuracy has involved hundreds of hours of observing time as well as countless more hours of analysis by the project's team of scientists. Their primary aim was to exploit the HST's superior optical acuity to resolve Cepheid variables in more distant galaxies, and so better calibrate the Tully-Fisher relation, supernovae luminosities, and other secondary distance indicators.

Their first efforts were thwarted by a flaw in the HST's primary mirror. But 10 years after this rather inauspicious beginning, the Key Project team announced in 2000 that the Hubble constant is $H_0 = 70\text{--}75$ km/s/Mpc with an uncertainty of 10 percent (see Figure 12.3).

Not everyone agrees, however. A lower value of 58 km/s/Mpc has been found by a team led by Allan Sandage, using the HST's Cepheid data to calibrate Type Ia supernovae (where a white dwarf in a close binary accretes sufficient gas from its partner to exceed the critical threshold mass of 1.4 Suns, thereby setting off a supernova explosion). An even lower value of 50 km/s/Mpc is claimed by investigators of gravitational lensing phenomena, where foreground galaxies gravitationally distort the light from more distant quasars. The lensing geometry yields the relative distances of the galaxy and quasar, which can then be related to the observed redshifts. Other astronomers, scrutinizing the "granularity" of elliptical galaxy images as a function of distance, find a higher value of 77 km/s/Mpc. So, despite the sterling performance of the refurbished HST and the indefatigable efforts of its Key Project team members, we are left with a Hubble constant that still ranges from 50 to 77 km/s/Mpc.

Since Hubble first proved that galaxies span incredible stretches of space and time, some of the boldest minds in astronomy have attempted to gauge the extent of the Universe. Rather than being discouraged that the scale of distances is still a matter of dispute, we should probably be grateful that nature has made it possible for us to find methods to measure a cosmos so unimaginably big. In the next chapter, we will make use of galaxy redshifts and the Hubble relation to map out the structure of the galaxian Universe within a few billion light-years of our vantage point. And in the following chapter, we will explore the universal expansion itself and the dynamics implied by its history.

13

CLUSTERS AND SUPERCLUSTERS, FILAMENTS AND VOIDS

A careful look at almost any portion of the sky with a large telescope will reveal a number of faint galaxies arranged in groups, all near the limit of visibility (see Figure 13.1). From the observed redshifts of these groupings, and the application of Hubble's relation between redshift and distance, we can infer that they are actual clusters of distant galaxies. Many of the galaxy clusters are themselves part of larger constructs—including superclusters, filaments, sheets, and bubbles. Much can be learned about the early Universe and its subsequent evolution through the study of these vast galaxian conglomerations. They are a clue to the remarkable happenings that occurred back when matter first separated from radiation, beginning its grand organization into our Universe of material things.

Richness

Clusters of galaxies range in population from meager ones like the Local Group to gargantuan assemblages with many tens of thousands of galaxy members, like the cluster in Coma Berenices. Astronomers refer to the number of galaxies in a cluster as its richness. How rich a cluster appears depends very much on the power of the observation. If our Local Group were observed from some distant galaxy, a small telescope might reveal only a loose association of three galaxies (the Milky Way, M31, and M33). But if an alien astronomer took a long-exposure photograph of the group, more galaxies would begin to fill in the space: NGC 205, M32, and the Magellanic Clouds would show up. But only with a very powerful telescope would there be any chance of seeing the abundant dwarfs in our group, such as Sculptor or IC 1613. Therefore, when we observe a distant cluster of galaxies, our judgments of its richness are partly subjective, a function of its distance from us and the size of our telescope.

In an attempt to put all clusters on somewhat equal footing, astronomers describe the richness of a cluster in terms of the number of galaxies it contains within a certain brightness interval. For example, the great cataloguer of galaxy clusters, George Abell, determined the population of clusters according to the rule that one counted those galaxies within the two-magnitude interval fainter than the third brightest galaxy. He chose the third rather than the first to avoid problems that might be caused if the first (or second) brightest happened to be a bright foreground object.

Abell identified his clusters and cluster members from visual examinations of photographic plates as they were being taken with the 48-inch Schmidt telescope at Mount Palomar in the 1950s. Distances were not known for most galaxies when he was making his pioneering catalogue and so foreground galaxies could indeed have skewed his counts. By Abell's counting rule, galaxy clusters in his catalogue (which excluded small groups like the Local Group) range in richness from about 50 to 300 qualifying members. The total number of galaxies in these clusters, of course, is far larger than this, many thousands for most of the clusters Abell studied. The richness measure just skims the cream off the top. Nevertheless, it provides a quantitative means of evaluating the relative numbers of galaxies in clusters.

Types of Clusters

Richness is just one way in which galaxy clusters differ. Abell also noted that some clusters seemed much more regular than others, and he separated them into two classes according to the regularity of their shape. In recent years a number of different classification systems have been invented, but all basically recognize that the types of galaxies contained within a cluster (spiral, elliptical, irregular) seem to be related to the shape of the cluster. Those with a smooth, regular shape are mostly made up of elliptical and S0 galaxies, while loose, irregular clusters contain many spirals and irregular galaxies.

The Coma cluster is a spectacular example of a regular cluster (see Figure 13.1 and Plate 28). It is compact and concentrated to its center, where there are some very luminous giant galaxies, NGC 4889 being the brightest. This

FIGURE 13.1 (facing page) Comparison of the rich Coma cluster of galaxies (top) and the less populated Virgo cluster (bottom). Both images were taken with the 4-meter telescope at Kitt Peak National Observatory in Arizona. The view of the Coma cluster (top) spans about 1.5 million light-years, but includes only the cluster's central half. Almost every object in the image is a galaxy. Elliptical and lenticular (S0) galaxies prevail, with the giant ellipticals NGC 4889 and NGC 4874 dominating. Only about one-seventh of the Virgo cluster is shown (bottom) in this million light-year closeup. This cluster shows a sparser population of galaxies than does the Coma cluster, with spirals represented along with ellipticals. The giant ellipticals on the right are M84 and M86.

elliptical galaxy is almost large and luminous enough to be called a cD galaxy, a classification reserved for certain extremely dominant giant galaxies that inhabit the centers of regular clusters and that are often found to be radio sources. We'll return to these supergiant galaxies and their remarkable story a little later in this chapter.

The Coma cluster has very few spiral galaxy members, and most of these are located in the outskirts of the cluster. The central concentration of ellipticals is high, with the density falling off steeply away from the center. Like many regular clusters, Coma emits radio radiation. Some clusters of its type are remarkably strong emitters of radio waves, originating from a few large, active galaxies as well as from a general background of intergalactic gas. We can even "see" this gas in some clusters; as radio galaxies plow through the gas, they leave jets bent around behind them, like tails trailing in the wind (see 3C 295 in Figure 11.2).

About a third of all regular clusters emit X-rays, an indication that their intergalactic gas is very hot (see Abell 1367 in Figure 4.5). In the case of Abell 1367, the X-rays come from gas with a temperature of about 100 million Kelvins. One explanation for the X-ray–emitting gas is that it results from violent collisions between galaxies, which sweep out each other's interstellar gas, so that they look like gasless S0 galaxies. The gas from each is heated by the collision and left near the place of collision. These forced exhalations mix with gas from other collisions in the cluster to form a hot intercluster medium.

The density of regular clusters is high enough, especially near their centers, that such collisions can happen, though not everyone agrees that they can happen often enough to explain the concentration of S0 galaxies in regular clusters. Walter Baade and Lyman Spitzer first proposed this idea back in 1951, when the distance scale was thought to be much smaller than it is now (see Chapter 12). With the new scale, the galaxies in a cluster are farther apart and can collide less frequently than the calculations originally indicated. Perhaps the collisions mostly occurred at an early state of the Universe, when galaxies were just forming and the Universe was denser. In any case, it is true that an unusually large number of S0 galaxies are found near the centers of regular clusters, with a corresponding lack of spirals.

Irregular clusters, on the other hand, contain lots of spiral galaxies. These clusters have an open sprawling structure, with little central concentration. Only about a quarter of them emit radio waves and less than 10 percent are X-ray sources. Collisions between members are expected to be rare enough that few S0 galaxies are created by the sweeping mechanism that is thought to occur in regular clusters. There is less hot gas as well, and these clusters are therefore rather quiet at radio and X-ray wavelengths.

The Virgo cluster is the nearest well-populated "rich" cluster and a good example of the irregular type. Spread out over several degrees in the sky, it contains many large, glorious spiral members, such as M100 and M61. It is so sprawling that its center is hard to locate without carefully counting members. There are subclusters within it and many small groups surrounding it out to great distances. Our Local Group is apparently an outlying suburb of the greater Virgo system.

To continue the urban analogy, the different kinds of galaxy clusters can be thought of as similar to different kinds of cities. Regular clusters are like New York, with a strong concentration to the central city, Manhattan, where one finds many giant buildings, all looking more or less alike. Irregular galaxies are more like Los Angeles, sprawled over the countryside with little central concentration. The middle area does contain a few large buildings with a variety of shapes, but there are other places miles away that also have local population centers. It is difficult to decide exactly where Los Angeles ends; it just slowly peters out at great distances, where there are still a few concentrations that may belong to the central city, though people will argue the point.

We live in a poor, irregular group of galaxies whose members are not especially obvious to us when we look out at the depths of space. What would it be like to live inside one of the great clusters like Coma? The density of galaxies near the middle of one of these is quite high, about 3,000 galaxies per cubic megaparsec (Mpc) for the top 8 magnitudes. The top 8 magnitudes includes all the giant galaxies and extends in brightness down to galaxies as faint as the irregular galaxy NGC 6822, but would not include extreme dwarfs like Sculptor. The galaxies would be packed together with an average distance of about 150,000 light-years, roughly the distance between the Milky Way and the Magellanic Clouds. If we were in such a cluster, the night sky would be filled with hundreds of galaxies bright enough to be seen easily without a telescope. It would be a magnificent sight, indeed!

Cannibalism

We have seen that galaxy merging is endemic throughout the Universe. In dense clusters of galaxies, the merging of many lesser galaxies can lead to the formation of a colossal elliptical galaxy. This process most likely explains the giant cD galaxies found near the centers of some regular clusters. Nearby examples of cD galaxies include the giant ellipticals M87 and M84 in the Virgo cluster.

A cannibalistic galaxy that eats other galaxies would be expected to have an extended outer envelope, as cD galaxies do, and sometimes to have more than one nucleus, owing to incomplete digestion of its recent meals, as some galaxies do. It will be overweight, having swallowed other galaxies until its mass is greater than the mass of more normal galaxies, and it will reside near the center of a cluster, being its most massive member and thus dominating the motions of the galaxies. The violence of the meals should result in a hot plasma of gas visible as radio radiation. All these characteristics are seen in cD galaxies, and the evidence seems overwhelming that they are the result of this process. In its own way, the Milky Way is a cannibalistic galaxy, as its ongoing ingestion of the Sagittarius dwarf galaxy attests. But its level of gluttony pales next to the porkers found in the dense centers of rich clusters.

Superclusters

Several early students of galaxy clusters noticed a tendency of some clusters to be clustered together in even grander clusters. One impressive example is the way that nearby groups seem to congregate around the Virgo cluster, making up what has come to be called the local supercluster. When Abell examined his catalogue of clusters, he found evidence for this kind of superclustering. The groups of clusters that he found were roughly 100 million light-years across and included an average of six clusters.

Research done in recent years has corroborated the existence of superclustering. Various kinds of statistical tests are necessary to be sure of the reality of such clumps; it is remarkably easy to see clustering that is not there, so astronomers have exercised great care in that respect. Most tests indicate that clusters of galaxies are frequently clustered together in groups ranging in diameter from about 100 to about 1,000 million light-years. They have masses about 10^{16} times the mass of the Sun. Most are not round in outline, but long and narrow (see Figure 13.2).

The local supercluster, the only one found that has a significant concentration to its center, includes the Virgo, Centaurus, and Hydra clusters

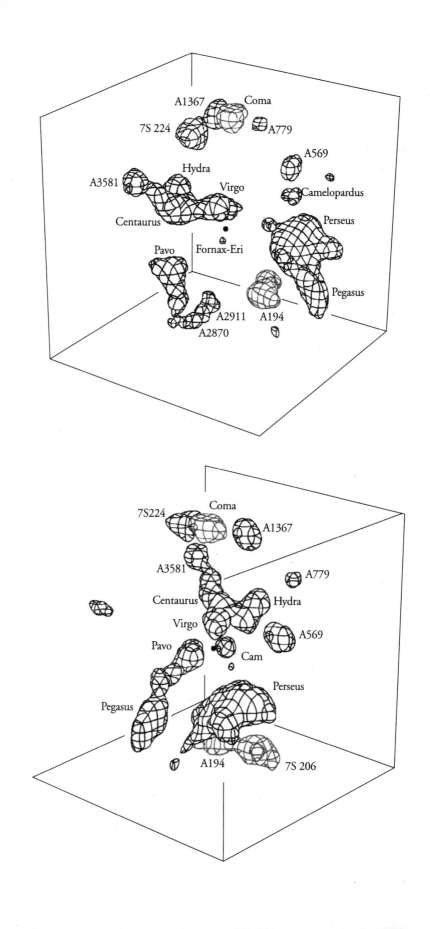

FIGURE 13.2 (facing page)   Two perspectives of the galaxy clusters within 80 Mpc (260 million light-years) of the Milky Way, represented here as a black dot in the middle of each diagram. The clusters are seen to group into superclusters. The bottom perspective shows them in rough alignment, delineating an even larger filamentary structure.

among others (see Figure 13.2). The other well-studied examples of super-clusters seem to have no particularly dense center, but are just loose group-ings of roughly equal clusters. The local supercluster's center is about 60 mil-lion light-years from us and its diameter is about 120 million light-years. It is flat, with the ratio of its major and minor axes being approximately six to one.

A much larger example is the Corona Borealis supercluster. The redshifts of its constituent galaxies indicate that it is about 1.5 billion light-years from us and has a diameter of over a billion light years. Indeed, it occupies much of the space that we can see when we look in the direction of that constel-lation.

Superclusters are rather different from star clusters or clusters of galaxies. For one thing, they are not dynamically well defined, since there has not been enough time in the life of the Universe for the subclusters to respond to each others' gravitational pull. The time it would take one cluster to cross the supercluster is typically about 300 billion years, many times the age of the Universe. That easily explains why superclusters do not have a regular struc-ture and are usually not concentrated to their centers, both conditions that result from gravitational action and mixing.

Filaments, Sheets, and Voids

On scales exceeding several hundred million light-years, the groups, clusters, and superclusters appear to congregate into even larger constructs (see Figure 13.3). Long narrow bridges of clusters, known as filaments, form intricate networks of luminous matter. These linear features may constitute the inter-sections of vast sheets. On the grandest observable scales, the organization of galaxies has been likened to a vast froth of bubbles—the bubble walls con-taining the galaxies, with the interiors being essentially devoid of luminous matter.

The origin of this hierarchical structuring continues to intrigue and baffle astronomers. Evidence seems to suggest that the observed inhomogeneity must have been in the works long before the galaxies themselves formed. Theories have been developed that track the emerging architecture from the

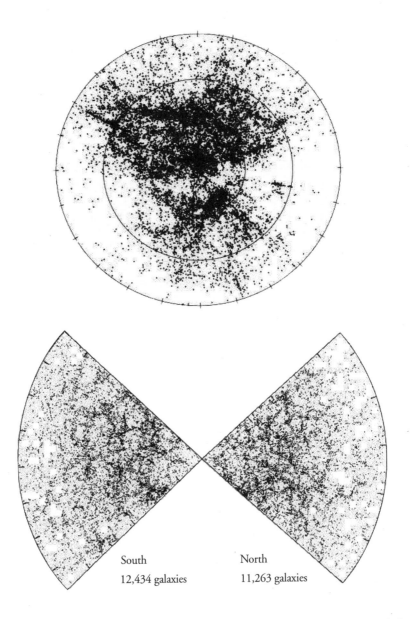

FIGURE 13.3 Galaxy distributions based on redshift surveys. Top: Puck-shaped diagram based on the Harvard-Smithsonian Center for Astrophysics redshift survey of galaxies within 200 Mpc (650 million light-years) of the Milky Way, which is located in the center of the puck. The local supercluster is the prominent concentration near the center. Above it, the Great Wall includes the Coma, Leo, and Hercules clusters in an elongated structure measuring 150 Mpc across. Below and slightly to the right, the Pisces-Perseus supercluster appears as a big clump. Bottom: Wedge diagrams based on a redshift survey of galaxies within 800 Mpc (2.6 billion light-years) of the Milky Way that was carried out at the Las Campanas Observatory in Chile. This much larger mapping shows galaxies in arrangements that suggest to some astronomers the presence of shell or weblike superstructures.

first density fluctuations that were present in the primordial Universe. Computer simulations—beginning with these sorts of "seed" fluctuations and relying on special blends of cold and hot dark matter—produce evolving pictures of clusters, superclusters, filaments, and voids that are remarkably similar to what we observe today (see Plate 31).

Perhaps soon, we will truly understand the sundry dynamics underlying the galaxian cobwebs that we see in the present-day Universe. Critical to our understanding is the expansion history of the Universe itself. In the next chapter, we will examine the evidence for cosmic expansion—and ponder the surprising trajectory implied by that evidence.

14

THE EXPANDING

COSMOS

If this book had been written just a few years earlier, it would have missed one of the most exciting discoveries of modern cosmology. Recent searches for distant supernovae have shown the host galaxies to lie at greater distances than would have been predicted from the galaxy redshifts and a freely expanding universe. At these greater distances, the galaxy redshifts appear anomalously low. And that means the early Universe—as traced by these remote galaxies—was expanding more slowly than it is today. Therefore, we appear to be living in a Universe whose expansion has accelerated over cosmic time.

Because classical gravity can only decelerate the expansion, these findings implicate some other force at work in the Universe—some sort of repulsive gravity that can drive the expansion to ever greater rates. Some astronomers attribute this force to a "dark energy" that pervades the cosmos. As we will see, this energy also vindicates the cosmological constant that first appeared in Albert Einstein's equation relating the various energies in the Universe. The cosmological constant was originally introduced by Einstein to counteract the contracting effects of gravity and so ensure a static Universe. Upon the discovery of universal expansion, Einstein dubbed the cosmological constant the biggest blunder of his career. Today, the cosmological constant not only is revived as a valid arbiter of the cosmic expansion, but is thought to embody most of the mass-energy in the Universe.

Before we grapple with these controversial findings, let us consider the basic observations and what they tell us about the expanding Universe.

The Expansion Paradigm

As discussed in Chapter 12, Edwin Hubble was the first to champion an expanding Universe. He based his claim on the observation that the spectra of distant galaxies were redshifted by amounts proportional to their distances. Interpreting the redshifts as velocity-induced Doppler shifts, Hubble established his now-eponymous relation $v_r = H_0 \times d$, which states that a galaxy's velocity of recession (v_r) increases linearly with its distance (d), the constant of proportionality (H_0) being known as the current-epoch Hubble constant.

Upon first consideration of the Hubble relation, one would be tempted to imagine the Universe of galaxies expanding away from our Milky Way galaxy—as if our Galaxy was somehow anointed with a special place of centrality in the cosmos. But the basic observations reveal that a far more comprehensive and democratic expansion is occurring. Such a universal expansion requires no center or edge. Sentient creatures in any other galaxy would observe the same relationship between redshift and distance that we have found with our telescopes here in the Milky Way. That is because the entire fabric of space is expanding, with all of the galaxies participating in the expansion like so many glittering sequins sewn into the fabric.

Imagine just such an elastic fabric studded with myriad sequins. Upon stretching the fabric, we might see that it is now twice its previous size. Similarly, the separations between the sequins are all doubled. If viewed from a particular sequin, the farthest sequins have receded the most—their displacement being equal to their initial distance. In similar fashion, the galaxies most distant from us appear to be receding the most—their redshifts increasing with distance according to the Hubble relation.

In many ways, the galaxy redshifts are tracing far more than the recession velocities embodied in the Hubble relation. That is because the expanding Universe stretches whatever light waves that traverse its reaches. Light waves from a distant galaxy have expanded with the Universe during the time that it took for the waves to travel from the galaxy to our telescopes. Therefore, the observed degree of wave stretching (or redshift) tells us directly how much the Universe has expanded since that light was first emitted by the galaxy. These sorts of considerations dispense with the somewhat misleading concept of recession velocity as it pertains to galaxy redshifts, replacing it with a more far-reaching paradigm that directly relates the redshift to the expansion of the Universe itself.

The Hubble Time

Stretching waves notwithstanding, Hubble's original relation provides a handy way to estimate the age of the Universe. If the galaxian Universe has been expanding at a constant rate, then one can backtrack the expansion to its origin—when all of the galaxies would have been as one. Rearranging the Hubble relation, one can obtain a timescale for this free expansion. The key is noting that the ratio of distance-to-velocity (d/v_r) has units of time, such that (kilometer)/(kilometer/second) = (second). Therefore, one can derive from the Hubble relation

$$T = d/v_r = 1/H_0.$$

Here, T is the Hubble time, an estimate for the age of the Universe that depends on the expansion neither accelerating nor decelerating over time.

In principle, one could express the Hubble constant, H_0, in units of (km/s) per (km) of distance, and so derive the age of the Universe in seconds. But astronomers typically choose to use those units that are most relevant to the observations. Favored units for the Hubble constant are (km/s) per (Mpc) of distance, with accepted values clustering around 60–80 km/s/Mpc (see Chapter 12). Taking the inverse of the Hubble constant yields a Hubble time, as measured in giga-years (1 Gyr = 10^9 years) of

$$T = 9.8 \ (100/H_0) \ \text{Gyr},$$

such that a Hubble constant of 70 km/s/Mpc results in a Hubble time of 14 Gyr. This is barely enough time to account for the ages of the most ancient globular clusters in our Milky Way galaxy, leaving only a billion or so years for the formation of the primeval Milky Way itself. As we shall see next, a Universe with a decelerating rate of expansion leads to even more problematic expansion ages.

Models of Expansion

In the discussion above, we have assumed that the Universe is in a state of "free" expansion, where the rate of expansion never changes. But our Universe contains lots of mass, and that mass should affect the expansion by virtue of its gravitation. Considerations of the mass in the Universe have led to a variety of models that have the expansion steadily decelerating over time.

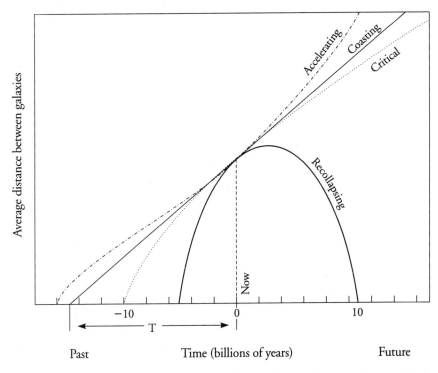

FIGURE 14.1   Space-time diagram showing the expanding trajectories of various kinds of universes. Time increases along the x axis, while the scale of the universe increases along the y axis. A freely expanding universe is plotted as a straight line. The timespan between its origin and the current epoch is the Hubble time (T). Universes with decelerating expansions have timespans that are shorter than (T), while universes with accelerating expansions have longer timespans.

These models have also enjoyed the cachet of relating to particular topologies of space-time.

The trajectory of any given universe can be handily visualized in a space-time diagram (see Figure 14.1). Here, cosmic time is plotted along the x axis, and the scale of the Universe is plotted along the y axis. The absolute size of the Universe need not be plotted; a scale relating to the mean separation between galaxy clusters, for example, will do just fine for our purposes. That way, we do not have to decide a priori whether or not the Universe is infinite in spatial extent—whew!

For a completely empty, freely expanding universe, the space-time diagram shows a straight line. The intersection of this line with the x axis indicates the moment of origin, from which we can readily measure the time to the present day—that is, the Hubble time. Topologically, such a universe would have "open" or "negative" curvature. The flaring surface of a saddle provides a two-dimensional analog to this type of geometry (see Figure 14.2). On such a surface, two initially parallel lines would ultimately diverge, and an inscribed triangle would contain angles whose sum falls short of 180°. Similarly, three mutually intersecting rays of light in an open universe would intersect at angles that add to less than 180°.

By adding matter to such a universe, one also adds gravity, which serves to brake the expansion. On the space-time diagram (Figure 14.1), the resulting trajectory has a slope that decreases with cosmic time—a graphical manifestation of the decelerating expansion. The intersection of this trajectory with the x axis now indicates an origin that is closer to the present day. Further addition of matter augments the gravity and the consequent deceleration, resulting in even shorter timespans between the origin and the current epoch.

If the matter density of the universe reaches a critical value, the resulting gravity is sufficient to slow the expansion and ultimately bring the universe to stasis. The space-time trajectory has a slope that decreases asymptotically to zero, so that the scale of the universe eventually stops growing. The intersection of the trajectory with the x axis is displaced from the current epoch by an amount precisely equal to two-thirds the Hubble time. If the current-epoch Hubble constant is 70 km/s/Mpc, then the expansion age of the Universe would be only 2/3 (14 Gyr), or 9.3 Gyr. With the ages of globular star clusters in the Milky Way exceeding 12 Gyr, this expansion age clearly violates the axiom that the Universe cannot be younger than the stars it contains.

When mass acts alone, such a critical universe has a "flat" or "zero" curvature, akin to the surface of a flat sheet of paper. As Euclid derived some 2,300 years ago, two initially parallel lines on a flat surface remain parallel, and the angles in a triangle add to precisely 180°. In our observed Universe, there is some evidence for light rays from the cosmic background having propagated through a similarly flat space (see Chapter 16). But we have yet to find sufficient mass in the Universe to achieve the critical density necessary for this sort of topology—which suggests that another form of mass-energy may be rampant.

Further addition of gravitating matter brings a universe past critical density and into the realm of "closed" curvatures. The space-time trajectory now shows two intersections with the x axis—one indicating its origin, and the other indicating its demise. Such a universe expands at a decelerating rate until it stops growing and begins to contract at an accelerating rate. The ultimate obliteration in a big crunch also leaves open the possibility of a big bounce, whereby a new space-time emerges from the collapse of the previous

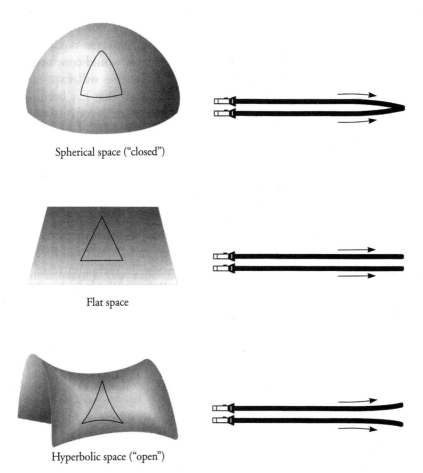

Spherical space ("closed")

Flat space

Hyperbolic space ("open")

FIGURE 14.2   The three basic types of cosmic topologies. Because we cannot picture curved three-dimensional space, curved surfaces must serve as our visual aids. A positively curved, "closed" universe can be likened to the surface of a sphere. On this surface, initially parallel lines converge, and the angles of a triangle sum to more than 180°. A "flat" universe of zero curvature is analogous to the surface of a flat piece of paper. Parallel lines in this planar geometry continue that way, while an inscribed triangle contains angles that sum to exactly 180°. A negatively curved, "open" universe can be likened to the surface of a saddle. Initially parallel lines diverge, while a triangle has lines intersecting at angles that sum to less than 180°. Recent angular measurements of small irregularities in the cosmic background radiation suggest that our Universe has a flat topology.

universe. Bounces could follow bounces in a cyclic pattern of emergence and destruction. As we will see, however, the material content of our Universe falls well short of supporting super-critical densities, while the recession rates of distant galaxies indicate an expansion that cannot be decelerating at the rate needed for eventual collapse.

Evidence aside, the topological aspects of a closed universe of positive curvature are fascinating to ponder. The surface of a sphere provides the two-dimensional analogue of such a space. A triangle drawn on this sort of spherical surface has angles whose sum exceed 180°, and by extrapolation, three mutually intersecting light rays would yield the same relation in a closed universe. Moreover, like the surface of a sphere, the volume of a closed universe is spatially finite—the only model to be so limited. In such a geometry, a light ray traversing the universe could ultimately return to its source. Therefore, a radio astronomer could send out a transmission in one direction and, given enough time, receive that same signal from the opposite direction—provided the universe does not collapse first. A galaxy observed in one direction could have ghost counterparts elsewhere in the sky. The mind boggles!

Evidence for Accelerating Expansion

To gauge the true trajectory of the Universe, it is necessary to compare expansion rates over cosmic time. This can be done by determining the redshift-distance relationship for galaxies at differing lookback times, and hence different epochs. The (relatively) easy part is getting the redshift. Thanks to large-aperture telescopes like the twin 10-meter Keck telescopes on Mauna Kea in Hawaii, obtaining high-quality spectra of galaxies at distances of several hundred million light-years is now fairly straightforward. The difficult part is determining the actual distance.

Until the 1990s, Cepheid variable stars were the best standard candles of well-known luminosity that could be used to fathom the distance to galaxies. Cepheids have led to reliable distances out to the Virgo cluster, some 50 million light-years away. Beyond Virgo, the Tully-Fisher relationship between a galaxy's absolute luminosity and rotation velocity has provided a secondary distance calibrator that is good out to about 300 million light-years and a corresponding lookback time of 300 million years. But one must go even farther out in space and further back in time to track any departures from free expansion.

Supernovae are the brightest sources of stellar origin, outshining the Cepheid supergiants by factors of a million or more. Consequently, they can be detected a thousand times farther away—in galaxies with lookback times of several giga-years. Supernovae of Type II, the explosive ends of massive stars, are too variable in output to be directly used as standard candles. But supernovae of Type Ia are well suited to the task. These occur in close binary systems, when a white dwarf star accretes from its giant partner

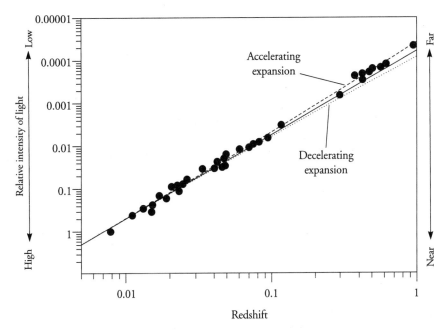

FIGURE 14.3   Hubble diagram containing both nearby and distant galaxies that have hosted Type Ia supernovae. At the greater distances and correspondingly greater lookback times, the plotted points deviate from the lower line, which traces decelerating expansion, and perhaps even from the middle line, which traces free expansion. The inferred accelerating expansion suggests that some sort of "dark energy" is extant whose pressure counteracts the decelerating effects of gravitating matter in our Universe.

enough mass to exceed the so-called Chandresekhar limit of 1.4 Suns, above which the white dwarf cannot support itself against the overwhelming gravity. At this critical moment, the white dwarf suddenly fails, the repercussions igniting a titanic supernova explosion that obliterates the remnant. Because the exploding mass is always close to 1.4 Suns, the explosion's radiant luminosity is remarkably constant—a standard candle indeed. Further insight on the luminosities of type Ia supernovae can be garnered from the rate of dimming, with the more luminous supernovae dimming the slowest.

Concerted efforts to calibrate nearby supernova Ia explosions and to find counterparts in distant galaxies began to pay off in the late 1990s. The results are shown in Figure 14.3. For the closer supernovae, astronomers have plotted a Hubble diagram whose slope indicates a Hubble constant of 64 km/s/Mpc—differing somewhat from the Cepheid-based results found by the HST distance-scale team. When the most distant supernovae yet detected are added, a cursory examination of the Hubble diagram reveals very little deviation from a straight line of constant slope. In other words, a freely expanding Universe is within the realm of possibilities, despite the fact that the Universe contains matter and should therefore be decelerating.

An even more amazing result arises when one examines the highest-redshift data points more closely. A small deviation from free expansion can be seen in the form of slightly lower redshifts at the inferred distances. If this trend is real, then the galaxies with the greatest distances and corresponding lookback times were receding more slowly than expected from a freely expanding Universe. Therefore, the early Universe—as traced by these distant galaxies—was expanding more slowly than it is today. In other words, the Universe appears to be in a state of accelerating expansion.

So far, this incredible finding is based on just a few dozen remote galaxies that have hosted type Ia supernova explosions. It is possible that the light from these explosions has been obscured by galactic and intergalactic dust. If so, the measured fluxes might be underestimated and the corresponding distances overestimated—and that could possibly explain the deviation in the Hubble diagram. But the dust would have to affect each of the measurements by an amount that steadily increases with distance in order to create such a systematic offset. Further measurements of type Ia supernova outbursts at high redshift plus better ways to handle the effects of dust will be necessary to confirm this provocative result. Meanwhile, the implications of an accelerating expansion are certainly worth considering.

The Dark Energy

One immediate outcome of an accelerating expansion is that the age of the Universe can exceed the Hubble time. As shown in Figure 14.1, the space-time trajectory of an accelerating expansion has an "origin" that predates that of a freely expanding universe. Depending on the detailed history of expansion, the time difference can amount to billions of years, which is plenty of wiggle room for the emergence of large-scale structures, of galaxies making up these structures, and of stars within the galaxies.

An accelerating expansion also provides an observable consequence for Einstein's cosmological constant, a term in his original equation of state that relates the various energies in the Universe. Among these sundry energies, the cosmological constant provided a sort of pressure throughout the cosmos. In the context of an expanding universe, the cosmological constant has been revived as a valid counterpoint to classical gravity. If our Universe is, in fact, expanding at an accelerated rate, then the cosmological constant must be even more influential than the braking effects of gravity. Therefore, the cosmological constant must involve copious amounts of energy, and this energy must come from somewhere.

The Universe is full of energy—some in the form of gravitating matter, some associated with the curvature of space itself, and some associated with the cosmological constant. The sum of these energies should equal the energy embodied in the universal expansion. By manipulating this cosmic energy equation, cosmologists can determine which mix of energies best fits the available measurements.

The energy density of luminous matter turns out to be very small—less than 1 percent of that needed to balance the expansion and yield a Universe with zero curvature. Such a flat space is favored by inflationary theory (see Chapter 15) and by recent observations of the cosmic background radiation (see Chapter 16). Adding the ordinary forms of dark matter (such as protons and neutrons in nonluminous constructs) brings the total matter energy density up to about 5 percent or less of that expected in a geometrically flat Universe. Weakly interacting cold dark matter and hot dark matter are thought to bring the material tally up to about the 30–40 percent level. That leaves about 60–70 percent to be provided by some other form of energy, the most favored candidate currently being that embodied in the cosmological constant. This energy is completely dark, complementing the dark matter that seems to dominate the gravitating energy. The origin of the dark energy is completely unknown, but theorists currently look toward the so-called quantum vacuum, a vast reserve of potential energy, from which the Universe itself may have been spawned.

As a consequence of this new agent in the cosmos, geometry no longer dictates destiny. The introduction of dark energy enables a flat universe (like ours) to expand at ever greater rates instead of slowing down. Depending on the relative amounts of dark energy and dark matter, a geometrically closed universe could continue to expand ad infinitum, and, conversely, a geometrically open universe could collapse on itself. That our Universe appears to be both geometrically flat and exquisitely poised between free and accelerating expansion continues to amaze even the most jaded of cosmologists.

If these observational measurements and theoretical arguments withstand the test of time, then we live in a universe that is expanding ever faster, yet is relentlessly Euclidean in form, a space-time that can be tracked by virtue of its luminous galaxies, yet is awash with dark matter and energy—a cosmos whose dynamical and structural evolution is driven by agents that we may never directly detect. Like mariners in the night, we are essentially blind to the tidal dramas unfolding in the deepest reaches of our world—barely cognizant of the heaving horizon beyond, yet full of imaginings. In the following chapter, great leaps of imagination are once again summoned, as we come to grips with the very earliest epochs of our Universe.

15

S C E N A R I O S

O F   O R I G I N

Once upon a time, some 13–20 billion years ago, our Universe began in what has become known as the Hot Big Bang. This seemingly fanciful scenario of universal origin was regarded with considerable skepticism shortly after the discovery of universal expansion. But the accumulation of evidence over the past half century has all but nullified alternatives to the Hot Big Bang, while adding insightful anecdotes to the basic tale. To understand the success of the Hot Big Bang theory in postulating our Universe's origin and describing its early evolution, let us examine some of the evidence as it exists today.

What We Know

Cosmic Expansion

First, there is the cosmic expansion itself. Running the clock backward leads one to imagine a time when the Universe was incredibly dense and fantastically hot—what George Gamow called the "primeval fireball" in *The Creation of the Universe,* his 1952 popularization of the Big Bang theory. This sort of thought experiment also leads to an ultimate moment of infinite density, where space and time have yet to emerge as separate constructs. A rough estimate for the moment of creation is given by the Hubble time, currently reckoned to be about 14 billion years, a figure that matches other chronometers of age in the Universe, as we shall see below.

The only way to avoid such a radical beginning is to counter the expansion with new matter that fills in the gaps. Running the clock forward (or backward) then leads to a universe whose matter appears (or disappears) at a rate that can sustain a constant average density ad infinitum. Herman Bondi, Thomas Gold, and Fred Hoyle postulated this Steady State theory of the Universe in the 1950s as a viable alternative to the Big Bang theory and its singular origin for the Universe. As we shall see, other pieces of evidence have since relegated the Steady State theory to the status of a provocative but seriously flawed cosmology.

One particularly glaring flaw pertains to the ages of galaxies. According to the Steady State theory, galaxies should manifest a wide range of ages. Some should contain only young stars. Others, like our Milky Way, should harbor stars with ages spanning 0—14 giga-years. Still others should consist exclusively of stars that were born hundreds of giga-years ago. But research on nearby galaxies has shown that the wide range of ages predicted by the Steady State theory is simply not present. We are hard pressed to find a galaxy made exclusively of young stars, and we have not found a galaxy that contains stars with ages significantly exceeding 14 giga-years. Instead, we find stars 10–14 giga-years old stars in pretty much every galaxy that can be closely scrutinized (see Chapters 6–8). An expansion age of 13—20 giga-years provides a reasonable fit to the stellar ages that we see in galaxies, this good match in ages bolstering the viability of the Big Bang theory.

Olbers' Paradox

Second, there is the matter of Olbers' paradox, a conundrum attributed to the nineteenth-century Viennese astronomer Heinrich Olbers, but also considered by Thomas Digges in 1576, Johannes Kepler in 1610, and Edmond Halley in 1721. Olbers' paradox asks the simple question: "Why is the sky dark at night?" If the Universe were infinite in both space and time, then any sightline should ultimately intersect the surface of a star somewhere in a far-away galaxy. That should result in a "night" sky as bright as the surface of a typical star—like one all-encompassing Sun. The effect is very similar to that experienced in a forest, where one cannot see the full extent of the forest for the trees. If the trees were self-luminous like the stars in galaxies, one would be surrounded by a continuous band of light that would outshine any nearby campfire. The fact that our nights are fairly dark suggests that something is not quite right with the "forest" of galaxies in which we dwell.

Edgar Allan Poe was also aware of Olbers' paradox, writing in 1848 that "were the succession of stars endless, then the background of the sky would present us a uniform luminosity." He was also one of the first to propose its solution—by imagining a Universe of finite age and by considering the fact that light takes a finite amount of time to travel from the depths of space to us. Given a fixed amount of time for light to propagate, there will be a distance beyond which we cannot see anything, because there wasn't enough time for the light to get to us.

By limiting the age of the Universe to 15 billion years or so, we limit the lookback time to galaxies by about the same amount, and the corresponding radius of the luminous Universe to about 15 billion light-years. The number of galaxies within this volume (about 50 billion) is far less than that necessary to illuminate the night sky like the interior of a furnace or the surface of the Sun. Therefore, Olbers' paradox tells us that our original assumption of infinite time is most likely wrong. The Universe had a beginning.

The Cosmic Background Radiation

The third major piece of evidence was initially regarded as unwanted noise in radio receivers that were being developed for commercial uses by scientists at Bell Laboratories. This radio "hiss" showed no preference for the antenna's placement and hence for direction in the sky. After eliminating instrumental artifacts and even the effects of bird droppings on the antenna, the scientists concluded in 1964 that the radiation was of cosmic origin. Further investigation showed that this background radiation had a thermal spectrum with a corresponding temperature of about 2.7 Kelvins (see Figure 15.1).

The same sort of thermal radiation had been predicted from the Hot Big Bang theory, first by George Gamow in his seminal work during the 1940s, and 20 years later by scientists at Princeton seeking the telltale signature. Little did the Princeton scientists know that just a few miles away, a mysterious cosmic background was being detected and diagnosed by Arno Penzias and Robert Wilson at Bell Laboratories—a feat that earned them the 1978 Nobel prize in physics.

This cosmic microwave background radiation (or CMBR) is now interpreted as coming from the Universe when it changed from an ionized plasma phase to a neutral atomic state some 300,000 years after the Big Bang. Once

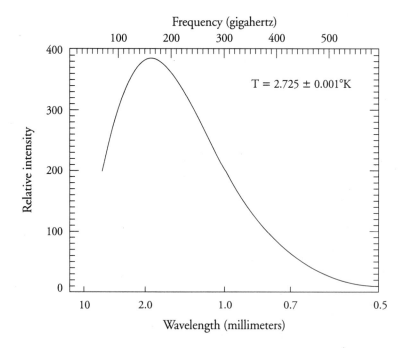

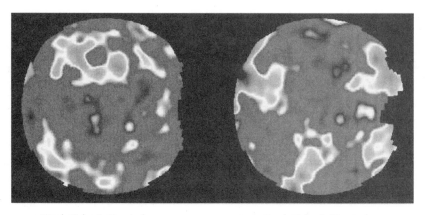

North Galactic Hemisphere   South Galactic Hemisphere

FIGURE 15.1 Top: The spectrum of the cosmic microwave background radiation (CMBR) was measured with exquisite accuracy in the early 1990s by the Cosmic Background Explorer (COBE). This spectral energy distribution was found to be nearly identical to that of an ideal black body radiator with a temperature of 2.725 Kelvins. Bottom: Variations in the intensity of the CMBR amount to no more than a few thousandths of a percent. These ripples in the firmament manifest the first hints of structuring in the primeval Universe.

the expanding Universe cooled enough for electrons to join with the protons, and so make atoms, the photons of light no longer had free charges to scatter off and were set free. Like the photosphere of a hot star, this "surface of last scattering" is what we see as the CMBR when we look back to the early hot Universe. The cosmic background radiation arose as visible photons from a Universe at a temperature of about 3,000 Kelvins, but has since been redshifted by a factor of a thousand, and has thereby been transformed into the microwave radiation that we measure today. By relating the CMBR to a particular epoch, the Hot Big Bang theory provides a cogent explanation for the CMBR as arising from a Universe that has undergone tremendous thermal evolution over cosmic time.

Primordial Nucleosynthesis

The fourth and final major buttress of the Hot Big Bang comes from the observed relative abundances of light elements in the Universe. During the first few minutes following the Big Bang, the Universe had a temperature and density similar to that in an exploding hydrogen bomb. And like a detonated H-bomb, the early Universe was busy forging hydrogen nuclei into heavier nuclei of deuterium, helium, and lithium.

By far, the major product was helium—amounting to 24 percent of the initial hydrogen—with the other elements adding to less than 1 percent. The exact amounts can be calculated on the basis of the temporal decline of temperature and density in the early Universe. From estimates of the current expansion rate and average density, one can backtrack to the first few minutes and so estimate the relative number of nucleosynthetic products that ensued. These numbers turn out to be remarkably similar to what is observed in stellar atmospheres and interstellar gas clouds. Once again, a thermally evolving (cooling) Universe is able to account for some of the most fundamental findings of observational cosmology (see Figure 15.2).

Further insights on conditions in the early Universe have benefited from mating the basic theory of the Hot Big Bang with elementary particle theory. In a synergistic relationship, particle physicists are informing cosmologists' notions of what the Universe was like during the first few seconds following the Big Bang. The cosmologists, in return, are elucidating how the early Universe can serve as the ultimate particle accelerator for testing theories of fundamental forces and particles. In the following chronology, some of the most transformative epochs of the early Universe are briefly summarized.

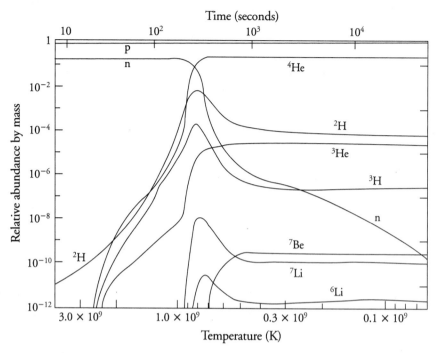

FIGURE 15.2 Nucleosynthesis in the early Universe. In a matter of minutes following the Big Bang, 25 percent of the hydrogen nuclei (protons [p]) fused with the available neutrons (n) to form helium-4 (^{4}He) nuclei and trace amounts of deuterium (^{2}H), helium-3 (^{3}He), and lithium (^{7}Li, ^{6}Li) nuclei. The good match between theoretical models (as shown here) and the observed relative abundances of light elements at the current epoch is regarded as one of the key successes of the Hot Big Bang theory.

A Brief History

Time ≅ 0 sec: The Beginning

That our Universe had a beginning some 13–20 billion years ago is as astonishing a discovery as any other in modern science. Yet the more we look at the evidence, the more we are compelled to imagine both space and time emerging from some sort of hyper–space-time—what some cosmologists today term the quantum vacuum. Otherwise, we are left pondering religious issues.

The quantum vacuum is likened to empty space, except that it is teeming with quantum fluctuations in energy, in accordance with Heisenberg's uncertainty principle. Formulated by Werner Heisenberg in 1927, at the dawning of modern atomic physics, the uncertainty principle states that it is impossible to know with infinite precision both a particle's position and its momentum—or a particle's energy at some infinitely precise time. At the beginning, both energy and time may have been fluctuating wildly. From this quantum chaos, submicroscopic worlds and antiworlds could have arisen, only to mutually annihilate in the quantum wink of an eye. Eventually, one of the virtual worlds may have survived the threat of annihilation, to become the Universe that we inhabit today.

To Einstein, the statistical nature of quantum theory's approach to matter was troubling, prompting him to complain that "God does not play dice with the Universe." Now, according to current thinking, God not only plays dice, but made our Universe from a lucky throw.

Time $\cong 10^{-43}$ sec, Temperature $\cong 10^{32}$ K: The Planck Epoch

The first chance we get to discuss the Big Bang in physical terms is after about 10^{-43} seconds have elapsed. Before this time, the density of the Universe would have exceeded 10^{94} grams per cubic centimeter—more than 10^{78} times the density of a proton! At such fantastic concentrations, the quantum nature of submicroscopic particles melds with the effects of gravity. In this quantum netherworld, our naive notions of space and time are scrambled beyond recognition. Until a viable theory of quantum gravity is developed, we are left speculating on this preemergent cosmos, where all forces were as one.

By the Planck epoch, named after the pioneering quantum physicist Max Planck, the expanding Universe had cooled enough for the gravitational force to "freeze out," and so work independently. Space and time began to have some meaning, and so our Universe began to take form.

The quantum particle that is thought to convey the gravitational force is known as the graviton. But no experiment has yet to confirm the existence of the graviton. Wavelike gravitational radiation is predicted from Einstein's theory of general relativity and has been detected indirectly (from the orbital decay of the binary neutron star system PSR 1913+16). Therefore, the gravitational fluctuations during the Planck epoch could have released similar gravitational radiation that has since permeated all of space-time. The prospect of actually detecting this radiation has helped motivate the development of gravitational wave telescopes, beginning with the twin Laser Interferometer Gravitational-Wave Observatories (LIGOs) that straddle flatlands in the moist forests of southeastern Louisiana and the desert plains of eastern Washington.

The remaining forces observed today—the strong force that binds nuclei, the weak force that governs nuclear fission, and the electromagnetic force that binds atoms—were still unified at the Planck epoch. The resulting behavior can be understood in the context of the grand unified theories (GUTs) of matter and energy, one of the holy grails of contemporary physics.

According to some GUTs, the most basic particles—quarks, antiquarks, leptons, and antileptons—can transmute into one another at sufficiently high temperatures and densities. GUTs also explain why electrical charges are quantized, while predicting that the minuscule neutrino has a finite mass, the proton decays over timescales of about 10^{32} years, and the magnetic monopole (a particle with one magnetic pole) should be ubiquitous. So far, experiments have yielded some support for neutrinos of finite mass, but not for the other major predictions of GUTs. Clearly, more work by both particle physicists and cosmologists will be necessary before we can understand the basic conditions that prevailed during the Planck epoch.

Time ≅ 10^{-35} sec, Temperature ≅ 10^{27} K): The Inflationary Epoch

After another 100 million Planck times, the Universe had cooled sufficiently for the strong force to break away from the weak and electromagnetic forces—these latter two still being unified under the so-called electroweak force. The emergence of the strong force involved a phase change in the Universe similar to the freezing of water to ice. Liquid water freezes into crystalline ice at a constant temperature by releasing its latent heat of fusion. Similarly, the freezing out of the strong force involved a tremendous release of latent energy. Like the Universe itself, this energy is thought to have come rather mysteriously from the quantum vacuum. By maintaining constant temperature and energy density in the expanding Universe, the quantum vacuum drove the expansion to ever greater rates.

The inflationary epoch is purported to have lasted only 10^{-33} seconds, but during this time, the Universe doubled in sized about 100 times—attaining a scale that was some 10^{30} times grander than it was previously. Before inflation, our observable Universe—out to a radius of 15 billion light-years—was no larger than the distance spanned by light in 10^{-35} seconds (about 10^{-25} centimeter, or one-trillionth the diameter of a proton). In this way, our preinflation Universe would have been in photon contact with itself and so could smooth out any egregious rough spots. Immediately after inflation, all that we can detect today would have fit inside the confines of a small town.

The inflationary epoch ended with another phase transition, as the latent energy was converted into a flood of photons and other particles. This initial "fireball" is what we see today in the cosmic background radiation, with the photon-particle "gas" having expanded and cooled to a temperature of 3,000 K, and the escaping radiation having since been redshifted by expansion to that of a black body at a temperature of 2.7 K.

Although much of inflationary theory appears outlandish and ad hoc, the theory is strongly favored by particle theorists, since it is based on the physics that underlie GUTs. It also handily solves several of the most intractable problems in Big Bang cosmology. By drastically inflating the Universe from a submicroscopic "seed," it ensures that the Universe was uniform in all directions (isotropic), smooth to a level of one part in 100,000, and essentially flat in curvature. All of these conditions are evident in the cosmic microwave background, and thus require some sort of explanation. Moreover, inflation theory predicts what kind of temperature fluctuations should be imprinted in the cosmic background radiation. Beginning with quantum fluctuations during the Planck epoch, inflation blows them up to become the thermal irregularities that are evident in the CMBR on the largest scales. No wonder cosmologists and particle physicists are so enamored with this seemingly bizarre theory.

Time $\cong$ 10^{-12} sec, Temperature $\cong$ 10^{15} K: The Particle Epoch

After a relative eternity (10^{11} inflation eras), the Universe had grown sufficiently cool for the electroweak force to bifurcate into the electromagnetic and weak forces. What was previously a hot broth of interchanging photons, W + Z bosons, neutrinos, electrons, quarks, and gluons could now settle down to become the sorts of ordinary matter that we have since inherited.

Protons, neutrons, and their antiparticle counterparts were the first to emerge from the cooling morass, their distinctive mass, charge, and spin states resulting from forged triplets of quarks. These relatively heavy particles are known as baryons, after the Greek "barus," which means heavy. Doublets of quarks became various kinds of mesons, which along with the baryon constitute the hadron family of elementary particles.

Members of the lighter lepton family—including electrons, neutrinos, and their antiparticles—had to wait for the Universe to cool for another 10^{-4} seconds before they could leave the bath of equilibrium reactions with other particles and become autonomous entities. Once the neutrinos "decoupled" from the matter, they could stream freely through space-time. In principle, we should be able to detect these primordial neutrinos. But the expansion of the Universe has since downgraded their energies by a factor of 10 billion, which makes any prospects of detecting them rather bleak.

The resulting particles and antiparticles are thought to have been equal in number to the photons. This situation could not last, however. An orgy of mutual annihilations among the particles and antiparticles led to the situation we have today, with one particle for every billion photons and nary an antiparticle to be found.

Time ≅ 100 sec, Temperature ≅ 10⁹ K: The Nuclear Epoch

After a couple of seconds had elapsed, the era of particle creation came to an end. By then, the Universe had cooled to below 10^{10} K, with no other particle transformations possible. For the next 3 minutes or so, the protons and neutrons fused together to make the nuclei of various light elements. Conditions were remarkably similar to those calculated to exist in the cores of supergiant stars just before they explode as supernovae. The thermonuclear explosion of a hydrogen bomb provides the most Earth-bound analogue, its behavior yielding essential benchmarks for understanding the cosmic nucleosynthesis reactions.

By the time the Universe had cooled to a few million degrees, 24 percent of the hydrogen nuclei (protons) had fused with the available neutrons to form helium-4. Along the way, trace amounts of deuterium (hydrogen-2), helium-3, and lithium-7 were made. The relative abundances of these elements can be observed today in stellar atmospheres and in the gas clouds filling interstellar and intergalactic space. The resulting agreement with the theory of Big Bang nucleosynthesis represents one of the most compelling bulwarks of the Hot Big Bang cosmology (see Figure 15.2).

The nucleosynthesis calculations further indicate a density of baryons that is only 2–5 percent of that necessary to achieve a topologically flat cosmos. If the baryon density were significantly higher or lower, the resulting mix of light elements would be measurably different from what we observe today. That the Universe appears to be geometrically flat strongly argues for other forms of matter-energy making their presence felt—yet another argument for nonbaryonic dark matter and dark energy pervading the cosmos.

Time ≅ 10¹¹ sec ≅ 10⁴ years, Temperature ≅ 10⁵ K: The Matter Epoch

Throughout the early history of the Universe, light dominated all other forms of energy. But as the Universe expanded, the wavelengths of this electromagnetic radiation were stretched and the corresponding photon energies were commensurately reduced. Therefore, the energy density of radiation decreased as the scale (R) times the volume ($4\pi R^3/3$) of the Universe; the result was a net $1/R^4$ decrease in radiation density. By contrast, the energy density of matter decreased simply as the volume, or as $1/R^3$. After 10^4 years, the matter density caught up with the radiation density and began to exceed it. From then until about the present epoch, the Universe and its dynamics have been dominated by matter.

Let us not forget the quantum vacuum, however. After its heyday during the period of inflation, it is thought to have permeated the Universe with residual dark energy. As the densities of matter and radiation have plummeted, this dark energy appears to have become ever more significant. Now, we find that the dark energy may account for two-thirds of the total energy density in the Universe—thus driving an accelerating expansion akin to the much more drastic inflation of yore (see Chapter 14).

Time ≅ 10^{13} sec ≅ 300,000 years, Temperature ≅ 3,000 K:
The Atomic Epoch

Not long after matter began to dominate the energy content of the Universe, the temperature had declined sufficiently for electrons to bind with nuclei and so form atoms. Once the temperature reached 3,000 K, the entire plasma of free nuclei and electrons had transformed to a neutrally charged atomic phase. For the photons, the sea of atoms presented a much more transparent medium of propagation. Instead of scattering off free charges every few atom crossings, the photons were set free to propagate throughout space-time.

This decoupling of the photons and matter also meant that the matter could finally congeal via its mutual gravitation without hindrance from the radiation field. Whatever irregularities had survived smoothing by the photons could now amplify, as the atoms in the Universe gravitated toward these primordial structures. Nonbaryonic dark matter had a head start, of course, having little if any interaction with the photons. Beginning with quantum fluctuations as early as the inflationary epoch, the dark matter would have been structured at a level of one part in 10^5 by the era of decoupling. The cosmic microwave background radiation manifests this tiny variation in density as an equally minuscule variation in the black-body temperature across the sky (see Plate 32).

Under the attractive influence of gravity, the over- and under-densities grew with the expanding Universe. Today, the Universe is a thousand times larger than it was in the decoupling era. The density inhomogeneities have grown proportionately, with the resulting condensations accounting for many of the structures that we can see today in the Universe—galaxies, galaxy clusters, superclusters, great walls, sheets, bubbles, and voids.

Lingering Questions

In the previous chronology of the Hot Big Bang Universe, we have emphasized the transformative processes that seem to best address the available observations. Yet before the first few minutes, relevant observations are sorely lacking. The actual form of the matter-energy in the very earliest Universe remains highly speculative. Grand unification theories predict certain types of particles (such as the elusive magnetic monopoles), while supersymmetry, superstring, supergravity, and other multidimensional theories predict their own menageries of primeval particles and forces. How these various "theories of everything" will fare in the face of new and better computations, experiments, and observations is at present completely unknown.

The observed predominance of matter over antimatter is but one empirical fact that begs for a more thorough explication. Once formed from photons, particles of matter and antimatter should eventually find one another and mutually annihilate back into photons. Because the particles are formed in pairs, their annihilation should result in a net zero sum of particles. Yet we live in a universe of matter, with antimatter virtually absent.

To create a universe with no antimatter, theorists postulate a very slight imbalance in the genesis of matter and antimatter particles. This tiny excess of matter (less than one part in a billion) is what survived after all the annihilations. How the initial imbalance arose, and how the annihilations proceeded during the earliest stages of the Universe remain lingering questions for cosmologists and particle physicists. Some constraints come from the observed ratio of photons to baryons (about a billion), suggesting that a billion annihilations occurred for every surviving baryon. But the theory that best accounts for this slight imbalance has yet to be identified.

Other major questions concern the genesis and nature of the dark matter and the dark energy. Much is made of the quantum vacuum as the ultimate provider, yet we have only a few tantalizing clues as to its presence from the behavior of certain laboratory experiments. The chaotic nature of the quantum vacuum also leads one to ponder the possibility that our Universe is not alone—that other space-times have spontaneously emerged from the vacuum and are evolving according to their own physical laws. Given these theoretical challenges, observations of deep space and early time continue to play a crucial role. In the next chapter, we will explore those observations that have worked wonders in improving our view of the cosmic frontier.

16

THE COSMIC
FRONTIER

Since the discovery of the cosmic microwave background, our knowledge of the early Universe has grown like wildfire. In some ways, we know more about the first 10,000 years—the so-called radiation era—than the subsequent billion years. Departures from thermal equilibrium were weaker during the radiation era, which makes the physics simpler, while the far-reaching effects of gravity were kept in check by the overwhelming torrent of high-energy photons. As the Universe continued to expand and cool, matter began to prevail over radiation. And after 300,000 years, the material Universe had cooled to a neutral atomic state that could decouple from the photons and so begin to congregate gravitationally. We see hints of this primeval structuring in the cosmic microwave background (see Figure 15.1).

After this critical epoch, however, a vast "dark age" confronts us. Lasting a billion or more years, this pregalactic period was characterized by cooling atomic gas, but with no stars to illuminate the night. The first starlight remains cloaked in mystery, for we have yet to build instruments that can clearly discern the highly redshifted stellar photons amidst the flood of photons from the epoch of decoupling. Tantalizing images of extremely faint sources at infrared and submillimeter wavelengths suggest that the dark age may not have been as long as once believed. But positive identification of these sources must await the next generation of space telescopes.

The first recognizable galaxies to loom on the cosmic frontier have photon energies and corresponding wavelengths that have been redshifted by about a factor of 5. That means we see them as they were approximately 12 billion years ago. Relating these faint blips at the edge of time to the grand swarms and pinwheels of stars that we find today remains a key challenge to astronomers. To see how far we have come, let's begin at the epoch of decoupling.

The Microwave Background

Upon its discovery in 1964, the cosmic microwave background radiation (CMBR) was little more than a faint radio hiss. Since then, concerted efforts have gone far in delineating both the spectral signature and spatial distribution of the CMBR. These efforts culminated in the launch of the Cosmic Background Explorer (COBE) satellite in 1989. Operating for a total of 5 years, COBE surveyed the entire sky at millimeter (microwave), submillimeter (far-infrared), and micrometer (mid- and near-infrared) wavelengths.

Spectral Purity

The millimeter-wavelength observations were most sensitive to the CMBR, whose emission reaches its maximum intensity in this wavelength regime. The spectrometer onboard COBE mapped out a spectral energy distribution that is indistinguishable from that of a perfect black body at a temperature of 2.73 Kelvins (see Figure 15.1). This exquisitely defined spectrum is the clarion call of the Hot Big Bang. Other, more local, explanations for the CMBR cannot readily yield such a perfectly monothermal spectrum without invoking ad hoc conditions. COBE's spectrum of the CMBR also places strong constraints on any primordial emitters or absorbers that may have existed shortly after the epoch of decoupling. Indeed, cosmologists are hard pressed to come up with primeval constructs of gas and stars whose unique contributions to the CMBR spectrum would have escaped detection by COBE.

Spatial Complexity

COBE was able to spatially map the CMBR at a resolution of 7 degrees, roughly 1/50 of a full circle. This sort of resolution is sorely insufficient for probing galaxy-size details but is adequate for mapping out irregularities in the CMBR on scales exceeding a billion light-years. The most obvious structure evident in the microwave sky was a dipole anisotropy that was approximately oriented toward and away from the Virgo galaxy cluster (see Plate

32). The deviation amounted to a 0.2 percent shift in the overall spectrum, the equivalent of a 600-kilometer-per-second Doppler shift toward Virgo.

We can interpret this effect as the result of several peculiar motions that our Galaxy is making in response to vast gravitating systems beyond the Local Group of galaxies. The nearest great attractor is the Virgo supercluster itself, of which the Local Group is a distant member. Our attraction to Virgo accounts for about 250 kilometers per second of the Doppler shift observed in the CMBR. The remaining deviation from pure universal expansion is attributed to another great attractor located about 65 degrees away from Virgo. Hiding behind the dust clouds of our Milky Way is a distant supercluster that is thought to contain 100,000 Milky Ways worth of gravitating mass. This is enough to move the whole Virgo supercluster by about 570 kilometers per second toward a Galactic longitude of 307° and latitude of 9°. Because the cosmic microwave background represents an inertial frame of reference that is independent of our special motion, we see it blueshifted in the direction of these great attractors and redshifted in the opposite direction.

After accounting for our anomalous motion with respect to the cosmic microwave background, COBE scientists obtained a map showing enhancements toward our Galactic midplane (see Plate 32). This time, the Milky Way was easily identified as the source of the excess emission. Thermal Brehmsstrahlung radiation from ionized gas and nonthermal synchrotron emission from high-speed electrons spiraling around lines of magnetic force are thought to provide the bulk of the observed Galactic features. Using measurements at much longer radio wavelengths, the scientists were able to calculate the Galactic contributions to the microwave mapping, and thereby remove them.

The remaining microwave background (shown in both Plate 32 and Figure 15.1) is thought to be solely of cosmic origin. Some of the lumps may be artifacts of the noise inherent to the residual emission, but the statistical ensemble of features provides a reliable measure of the large-scale irregularities in the CMBR. The average deviation from a perfectly monothermal background is remarkably low, amounting to no more than a few parts per 100,000. In mapping these minuscule deviations, COBE provided our first glimpse of how the early Universe was organized.

Primordial Fingerprints

The observed distribution of feature sizes in the sky indicates roughly equal amounts of structuring on scales ranging from 7° to 70°. These structures, at the epoch of decoupling, spanned distances so large that neither light nor any other conceivable interaction could have modified them in the allotted time. Yet they are all uniformly distributed around the sky, without any preference for location. Cosmologists explain such enormous and widespread features as the "fingerprints" of submicroscopic quantum fluctuations that emerged during or shortly after the Planck era and were subsequently blown up by a factor of 10^{30} during the inflationary epoch (see Chapter 15). Perhaps these thermal irregularities represent the seeds of the largest material structures that we can trace today in the distribution of galaxies (see Chapter 13). They may, however, also indicate gravity waves, gravitational lensing, sound waves, or other processes in the primordial Universe that could just as likely have left their mark on the CMBR.

Cosmoseismology

Further insights from the microwave background require measuring its structure on ever finer scales. This challenge is being met with a battery of microwave "telescopes" that observe from groundbased stations in Antarctica and aboard high-flying balloons—where microwave-absorbing water vapor is frozen out of the sky. To date, the BOOMERANG (Balloon Observations of Millimetric Extra-galactic Radiation and Geophysics) experiment has yielded some of the most exciting results. Operating at a resolution that is 10 times finer than that achieved by COBE, BOOMERANG mapped a 30°-wide patch of the southern sky. The resulting distribution of sizes in the mottled emission revealed a strong peak at sizes of about a degree. This sort of peak had been predicted by inflationary cosmologists who espouse a topologically flat Universe as a natural outcome of the drastic inflation.

According to most models of the Hot Big Bang, the Universe just before the epoch of decoupling was a hot "gas" of strongly interacting baryons and photons, through which soundlike pressure waves could propagate—compressing some parts slightly and thinning others out. Driving the sonic tumult were aggregates of weakly interacting dark matter, whose gravity could still attract the baryons and refract the photons. At the moment of decoupling, the ionized plasma became atomic and the photons were set free. This phase change in the gas squelched any further propagation of the acoustic waves, but left the final acoustic compressions imprinted in the CMBR.

The strongest acoustic oscillations had wavelengths corresponding to the distance propagated over the lifetime of the Universe, then about 300,000 years. One might liken this acoustic mode to the fundamental tone of a vibrating string. Depending on the specific mix of baryons and nonbaryonic dark matter, astrophysicists can calculate the absolute wavelength of this fundamental acoustic tone to be something like 200,000 light-years. The observed angular size of this acoustic mode then depends on the expansion factor of the Universe since the CMBR was first emitted (about 1,000), and on the curvature of the space-time through which the CMBR has propagated. A positively curved space-time will yield larger angular sizes than a flat, or negatively, curved universe (see Figure 14.2). BOOMERANG observed an acoustic peak at an angular scale of 1 degree, indicating a space-time that is essentially flat, and thus corroborating one of the key tenets of inflationary cosmology.

Before 2010, NASA's Microwave Anisotropy Probe (MAP) and the European Space Agency's Planck Surveyor will have mapped the full sky to an angular resolution of 0.1 degree. If successful, these missions will probe fluctuations in the CMBR on scales of 50 million to 500 million light-years (including the 1,000-times expansion factor), thereby diagnosing the precursors of galaxy clusters, superclusters, and great walls. Resolving these finer fluctuations will also help to pin down some of the most important cosmological parameters, including the current-epoch expansion rate (H_0), the deceleration (or acceleration) parameter (q), and the sundry densities of mass-energy that mediate the cosmic expansion (see Chapter 14).

The Infrared Background

Like the microwave background, the observed infrared background is actually a complex blend of various foregrounds along with whatever cosmic background may exist. To dig out the cosmic signal requires both superlative data and tremendous tenacity by the researchers. The infrared camera on board COBE had already produced stunning all-sky views of our Galaxy (see Plates 6 and 7). But these wondrous vistas would have to be eliminated in order to discern any residual emission from the distant and primeval Universe. By modeling first the zodiacal light from dust in our Solar System, then the collective sources of infrared light in the Milky Way, and finally sources that have been identified with known galaxies and quasars, COBE scientists were able to carefully strip away these foregrounds and so produce an all-sky map of the cosmic infrared background (see Figure 16.1).

The resulting pointillistic scene shows variations on scales of roughly half a degree. Some of these speckles could be newborn clusters of galaxies at high redshift. Alternatively, the infrared flecks could be tracing supernova explosions in infant, otherwise invisible galaxies. Both possibilities push back the epoch of rampant star formation to little more than a billion years after the

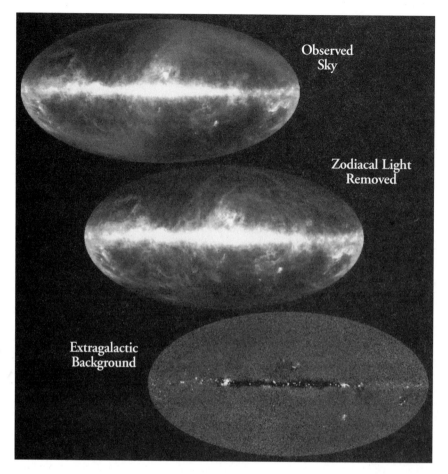

FIGURE 16.1   Top: All-sky mapping of the infrared sky, as surveyed by COBE's Diffuse Infrared Background Experiment (DIRBE). Middle: The same all-sky view after subtraction of the zodiacal light. Bottom: Residual emission after subtraction of contributions from our Galaxy and other identified galaxies and quasars. The remaining freckles of infrared light are thought to represent very distant and young galaxies and galaxy clusters.

Big Bang. So it seems that almost immediately following the epoch of decoupling, the galaxian Universe was beginning to take hold.

A higher-resolution view of the infrared sky was obtained by the Infrared Space Observatory (ISO) during its 1995–1998 deployment. By repeatedly staring at selected patches of apparently blank sky, ISO built up deep images that revealed a blizzard of extremely faint sources. Of the thousand or so faint objects, most have been attributed to galaxies at lookback times of 5 billion years or less. Several of the galaxies have optical counterparts. Of those, many are far brighter in the infrared, which most likely indicates that they contain enormous quantities of obscuring dust. These juvenile galaxies appear to be in the throes of titanic starburst activity akin to the ultra-luminous infrared galaxies (ULIRGs) observed in the local Universe (see Chapter 10).

The remaining faint sources in the ISO images have no optical counterparts. Either they are even more distant galaxies, or they are fragmentary protogalaxies, too far and faint to be fully resolved. Similar scenes have been recorded at far-infrared to submillimeter wavelengths with the Submillimetre Common-User Bolometer Array (SCUBA) camera on the James Clerk Maxwell Telescope (JCMT) atop Mauna Kea in Hawaii. Further revelations of the deep infrared sky are anticipated with the next generation of infrared space observatories, including NASA's Space Infrared Telescope Facility (SIRTF) and James Webb Space Telescope (JWST). At longer wavelengths, the European Space Agency's Far-Infrared Submillimetre Telescope (FIRST) and the international Atacama Large Millimeter Array (ALMA) promise to diagnose the very earliest epochs of galaxy formation by detecting galaxies and protogalaxies at redshifts of ten and higher.

Gamma-Ray Bursters

Bursts of cosmic gamma-ray emission have been recorded since 1967, when the first military satellites were deployed to monitor aboveground atomic bomb blasts. For 30 years, the nature of these high-energy phenomena remained a mystery. Even after the deployment of NASA's Compton Gamma-Ray Observatory (CGRO) in 1991, there remained two basic schools of thought. Either the gamma-ray bursts were coming from convulsing neutron stars in the halo of our own Galaxy, or they were incredibly more powerful outbursts from far more distant galaxies. The uniform (isotropic) spatial distribution of bursters provided the first hint that they did not belong to our Galaxy (see Figure 16.2). The major breakthrough came when the first positive identification of a gamma-ray burster was achieved in 1997.

Although the detectors onboard the CGRO were sensitive to most gamma-ray emission, the "optics" were crude at best. The resulting spatial resolution was little better than a few degrees on the sky. To do better, gamma-ray scientists had to rely on follow-up observations at other wavelengths. Their dreams were realized when the Italian-Dutch BeppoSAX X-ray Observatory was launched in 1996. A year later, BeppoSAX caught one of CGRO's gamma-ray bursts in progress, pinpointing its location to within a few arcminutes, the equivalent resolution of the human eye. Follow-up im-

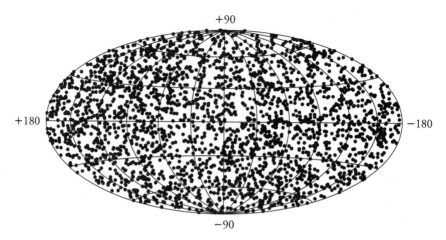

+90

+180

−180

−90

FIGURE 16.2 The spatial distribution of gamma-ray bursts in the sky. As in the other all-sky mappings in this book, the long axis is associated with the Galactic midplane, and the center is where the Galactic bulge and nucleus are located. The gamma-ray bursts show no preference for any Galactic feature, this mismatch suggesting that most of the bursts are not of Galactic origin.

aging with the Kitt Peak 0.9-meter telescope further refined the position to within an arcsecond, and optical spectroscopy with the Keck I 10-meter telescope revealed light from a galaxy at a redshift of 0.83, placing the burster more than 6 billion light-years away from our Galaxy.

An even more powerful burster was recorded 6 months later. In this burst event, the Hubble Space Telescope was able to focus on the position provided by BeppoSAX and image the optical afterglow along with the faint host galaxy. Deep spectroscopy with the Keck II telescope showed the galaxy to have a staggering redshift of 3.4, with a corresponding distance of at least 10 billion light-years.

Given the observed flux of gamma-ray photons and the distance inferred from the host galaxy's redshift, scientists estimate that the 20-second-long outburst had a luminosity the equivalent of 100 million Milky Way galaxies—more powerful than anything else known in the Universe. If this burst event and subsequent other confirmed gamma-ray bursts are representative of some new class of ultra-luminous object, then we might be able to use them to probe the very early Universe.

As some astronomers believe, the gamma-ray bursters could be extreme versions of supernova explosions—hypernovae—in which the most massive stars imaginable end their lives by imploding into black holes. Such stars are thought to have populated the Universe shortly after the decoupling epoch, and perhaps before the first galaxies were assembled. It is possible that we will be able to witness the explosive deaths of these remote stars in the form of gamma-ray, X-ray, and visible outbursts. Not surprisingly, the next generation of gamma-ray and X-ray satellites are being developed in close coordination with the largest optical telescopes to seek the telltale signatures from these first-generation hyperbombs.

The View from Hubble

After more than 20 years of gestation, the Hubble Space Telescope (HST) was released from the cargo bay of the space shuttle *Discovery* in April 1990. From the outset, astronomers intended to achieve two major goals with this great observatory: (1) to pin down the distances of galaxies out to the Virgo cluster, and thereby constrain the Hubble constant to better than 10 percent accuracy; and (2) to probe the depths of space with unsurpassed sensitivity and resolution, thereby pushing back the veil of time to the epoch of galaxy formation.

Unfortunately, these lofty goals had to be shelved shortly after first light, when the telescope made its first astronomical observations, and the HST astronomers found signs of severe spherical aberration in the resulting images. To their horror, they discovered that the HST's primary mirror had been figured—with exquisite precision—to the wrong specification. Another 3 years would pass before a repair mission would outfit the HST with instrumentation whose optics could compensate for the aberrant primary.

The Medium Deep Survey

Hobbled but undeterred, astronomers made the best of those years by implementing the HST Medium Deep Survey. While HST's spectrograph or photometer was fixed on some specified target, the Medium Deep Survey team used the camera in parallel mode to image adjoining pieces of the sky. Through this parallel program, they were able to build up a catalogue of 120,000 galaxies out to a redshift of about 1 and a corresponding lookback time of about 7 giga-years.

Somewhat surprisingly, the giant elliptical and spiral galaxies did not appear much different at the earlier epochs. Much of the morphological changes could be explained by the large distances and redshifts of these adolescent systems. Follow-up photometry suggests that the giant galaxies were significantly brighter in the past, however, perhaps because of greater levels of starbirth activity having occurred back then. The greatest case for evolu-

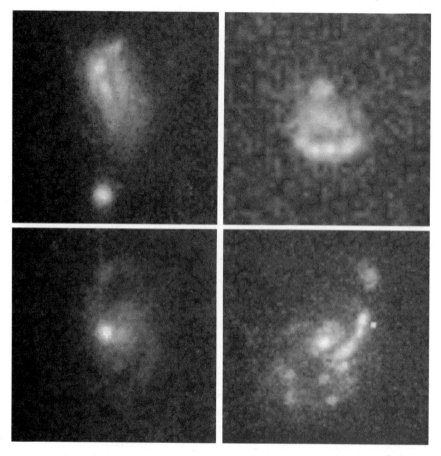

FIGURE 16.3   The HST Medium Deep Survey found a large number of faint blue galaxies at lookback times of a few billion years. They appear to be starbursting irregular galaxies but may be ultraviolet-bright parts of larger spiral galaxies.

tion over cosmic time was found with a class of faint blue galaxies whose ragged shapes resemble the Magellanic irregular galaxies observed in the Local Group (see Figure 16.3). The detected numbers of these galaxies show a strong rise at greater redshift, which suggests that they were more abundant around 7 giga-years ago.

Perhaps the faint blue galaxies were more numerous at earlier epochs and have since been absorbed by their larger neighbors through merging. It is also likely, however, that we are seeing an evolution in luminosity rather than numbers, making these objects less evident at later epochs and correspondingly smaller redshifts. Many of the faint blue galaxies may also represent starbursting portions of much larger galaxies, whose ultraviolet emission

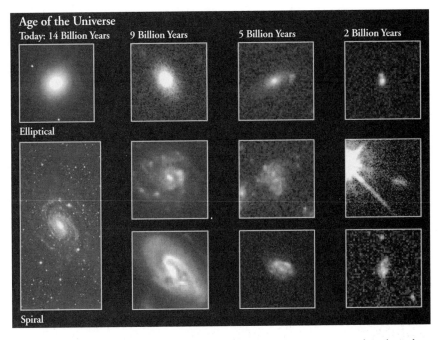

FIGURE 16.4 A chronological sequencing of galaxies over cosmic time based on deep observations with the Hubble Space Telescope. Each sequence is segregated according to Hubble type. At the earlier epochs, the galaxies appear less distinct and symmetric. This may be a real evolutionary effect, or it may be the consequence of having imaged the young galaxies at ultraviolet wavelengths that have been redshifted into the visible spectrum. Because ultraviolet emission arises primarily from starbursting regions, the high-redshift images are biased toward the most active star-forming regions in the galaxies.

has been redshifted into the blue part of the spectrum. If so, we are not seeing the emergence and dissolution of a separate class of galaxies, but instead are witnessing the rise and fall of starburst activity in giant galaxies (see Figure 16.4).

The Hubble Deep Fields

To discriminate among these possibilities, it would be necessary to look even more deeply into space and time. Toward these ends, the fully refurbished HST was pointed at a tiny patch of sky near the Big Dipper for a total of 10 days in 1995. The resulting "core sample" of the galaxian Universe revealed a plethora of weird-looking realms all the way down to the 30th-magnitude detection limit (see Plate 30). This image of the so-called Hubble Deep Field (HDF) is today the most famous image yet produced by HST. Together with its Southern Hemisphere counterpart (an image taken in 1999), the HDF represents our deepest view yet of the galaxian Universe.

By imaging the northern and southern HDFs through different color filters, astronomers could render color pictures that revealed the full range of galaxy types. Even more important scientifically, they could determine the approximate wavelength at which a galaxy's light dims into the background. The agent responsible for dimming the galaxy's light is galactic and intergalactic atomic hydrogen, which absorbs ultraviolet photons having rest wavelengths of less than 912 Angstroms—the Lyman limit, named after the atomic physicist Theodore Lyman. By comparing the observed wavelength of dimming with the rest wavelength of the Lyman limit, astronomers could calculate a redshift for each galaxy. Follow-up spectroscopy with the Keck telescopes verified this photometric method for obtaining redshifts.

The comparative ease of the photometric method has enabled hundreds of redshifts to be determined from the HDF images. Although the more prominent galaxies show redshifts less than 1, the fainter, smaller, and redder galaxies have redshifts of up to 5, with corresponding lookback times of 12 giga-years. The recorded light from these faint blips was originally emitted at ultraviolet wavelengths, and has since been redshifted into the visible range. Therefore, we must exercise caution when trying to interpret what we see. While some astronomers interpret the linear-shaped galaxies as primordial "chain galaxies," other astronomers attribute the linear features to relatively normal disk galaxies as seen edge-on. Similar disagreements prevail over the perceived lack of central bars and bulges, as well as the seemingly lumpy nature of the observed emission.

Are we seeing subgalactic clumps prior to assembly, or highly redshifted ultraviolet emission from discrete starburst regions that reside in much older, larger, and more regularly shaped galaxies? And what about the effects of close interactions? Do they dominate the organizing dynamics and resulting morphologies at large lookback times? Having pondered the marvels revealed in the Hubble Deep Fields, we find ourselves at new thresholds of understanding—and questioning.

As the Medium Deep Survey showed photometric variations over the last 7 giga-years, the Hubble Deep Field reveals an evolution in the total starlight that spans the last 12 giga-years. The ultraviolet luminosities indicate starbirth rates that were much higher than those of normal disk galaxies at the current epoch. Indeed, the density of star-forming activity throughout the Universe appears to have reached a maximum at redshifts of about 3 and a corresponding lookback time of about 10 giga-years. Subsequent studies of the obscuring effects by dust suggest even higher rates of star formation going all the way back to redshifts of 5 or more, when the Universe was only 2 giga-years old. These sorts of analyses show that the transformation of gas to stars was at its peak in the early galaxian Universe and has since declined to less than a tenth of its previous heights.

Whether the galaxies were fully formed at large lookback times remains a question that must await high-resolution imaging at longer wavelengths. Some deep imaging with HST's infrared camera has revealed disks, bulges, and bars of cooler and presumably older stars underlying the starburst activity. But a more thorough and sensitive survey at infrared wavelengths will be required to delineate the types of galaxies that first populated the Universe.

The Intergalactic Medium

Between the galaxies are vast constructs of extremely tenuous gas. Thanks to the brilliant and remote quasars, we can probe this gas as it absorbs the background quasar light according to its chemical composition and thermal state. What we find in deep spectra of distant quasars are entire "forests" of absorption lines associated with the foreground gas clouds (see Figure 16.5). The Lyman-alpha forest, for example, consists of absorptions by hydrogen as it changes its energy from the groundstate to the first excited level. Each absorption line (or "tree") in the Lyman-alpha forest then represents a separate cloud at a particular redshift.

By studying the distribution of redshifts, astronomers have been able to piece together the distribution of absorbing matter—all the way out to redshifts of 5 or more. Their findings reveal an intergalactic medium that has evolved drastically over cosmic time. What was a cloudy haze of hydrogen some 12 giga-years ago has congealed into a tangled web of discrete filaments, sheets, and voids—much like the large-scale distribution of galaxies in the local Universe. This structuring of the gas appears to have taken only a few billion years to occur. Computer simulations vivify the evolution implied by the absorption-line studies (see Plate 31), further corroborating the observations and their implications.

There are other groups of spectral absorption lines that can be used to probe the intergalactic medium, including the ultraviolet absorption-line systems of ionized magnesium and carbon. The strengths of these particular absorptions indicate the degree of chemical processing into and out of stars that has occurred within the cosmic time allotted. Astronomers find that the processing and its widespread distribution must have begun very early in the history of the Universe, because both the quasars and the absorbing clouds at high redshift are enriched with heavy elements.

Recent observations of quasars with HST and the Hopkins Ultraviolet Telescope (HUT) have detected the long-predicted absorption signature

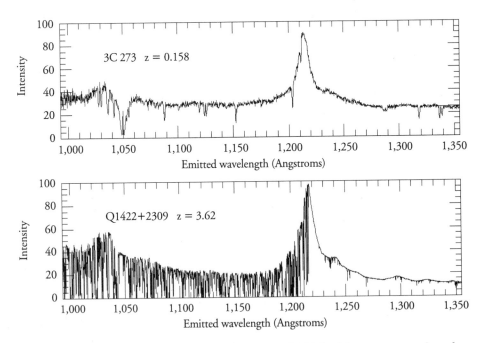

FIGURE 16.5   The spectrum of the nearby quasar 3C 273 (top) has a sparse number of Lyman-alpha absorption lines compared to the "forest" of lines that is evident in the spectrum of the highly redshifted and hence remote quasar Q1422+2309 (bottom). Both spectra show the Lyman-alpha emission line at its rest wavelength of 1,216 Angstroms, after the disparate redshifts (z) of the quasars have been compensated for. Each absorption line at shorter wavelength is produced by a different cloud of hydrogen along the line of sight to the quasar. The disproportionately large number of blended absorbers in the spectrum of the more distant quasar shows that the intergalactic medium was more congested at the earlier epochs traced by the remote quasar.

from intergalactic ionized helium. Because the helium is ionized, it is probably associated with copious quantities of ionized hydrogen—a virtually undetectable but major component of the intergalactic medium. How the hydrogen and helium got ionized remains unknown, but theorists are considering various kinds of energetic phenomena early in the history of the galaxian Universe (including super-massive stars and primordial explosions) that either irradiated or shocked the gas back to a plasma state.

Gravitational Lensing

The bending of light by gravity was predicted by Albert Einstein in 1911 and verified by Sir Arthur Eddington during the total solar eclipse of 1919.

By precisely measuring the angular displacement of stars behind the eclipsed Sun, Eddington and his colleagues showed that the starlight had been deflected by the Sun's gravity in exact accordance with Einstein's theory of general relativity.

On galaxian scales, we can observe multiple images of distant quasars whose light has been gravitationally lensed (bent) by individual foreground galaxies. We can also detect multiple images of fainter galaxies that have been magnified and distorted by the strong gravity of intervening galaxy clusters (see Plate 29). The resulting optical "mirages" speak volumes about the remote sources, the masses responsible for the gravitational lensing, and the intervening geometry of space.

For lensed quasars, the variable output of the quasars can be exploited to pin down the distances of both the quasar and the lensing galaxy. Off-axis images of the quasar will fluctuate out of phase with the central on-axis image according to differences in the respective paths traveled by the different light rays. By monitoring the phase differences and applying some basic geometry, astronomers can derive the distances to the quasar and lensing source. In this way, the Hubble relation between distance and redshift can be extended to much greater distances, thus yielding stronger constraints on the Hubble constant and other cosmological parameters. To date, such investigations have indicated a Hubble constant of about 50–65 km/s/Mpc, somewhat smaller than the Cepheid-based value of 70–75 km/s/Mpc.

For galaxies lensed by foreground clusters, the distribution and geometry of the multiple images reveal the nature of both the lensing mass and the lensed source. In the case of the Pretzel galaxy, shown in Plate 29, the galaxy cluster CL0024+1654 has produced five replica images of the background galaxy. The degree of displacement and distortion of the individual images has been used as input for modeling the distribution of mass in the foreground cluster and for reconstructing the background source.

The iterative modeling settled on a solution that includes overwhelming quantities of dark matter mixed in with the clustered galaxies (see Figure 4.4). The corresponding reconstruction of the magnified and distorted background source has all the earmarks of a disk galaxy with a star-forming ring (see Plate 29). At a redshift of about 1.7, the so-called Pretzel galaxy attests to the efficacy of disk formation at early epochs.

The strong lensing exerted by galaxy clusters such as CL0024+1654 provides powerful vindication of the existence of dark matter in the Universe. Elsewhere in the cosmos, vast blobs of unseen matter are probably imparting weaker distortions on the distribution and shapes of distant galaxies. This so-called weak lensing has been sought through deep electronic imaging of wide angular fields. The numerous faint blue galaxies at distances of 3 to 6 billion light-years are especially well suited to this sort of tomographic study. A statistical analysis of the distortions evident in the galaxies has yielded the first images of the broadly distributed dark matter. The preliminary maps are reminiscent of the filamentary structure seen in the large-scale distribution of luminous galaxies.

Perhaps the most ambitious application of gravitational lensing is as a diagnostic for determining the geometry and size of the Universe. If the Universe is flat, and if the cosmological constant is driving an accelerating expansion, then the most distant galaxies are considerably farther for their redshifts than in a decelerating cosmos without a cosmological constant (see Figure 14.3). In such a larger Universe, there will be greater amounts of gravitating matter along any line of sight, and hence a greater lensing effect on the most distant galaxies. This too can be tested through deep wide-field imaging. So far, the results have been inconclusive.

Like the illusions in a far-flung hall of mirrors, the sundry effects of gravitational lensing are tricky to decipher. Yet the payoffs in our understanding of the Universe continue to make gravitational lensing a compelling topic of study.

Our Galaxian Legacy

The most complex and comprehensive data on the cosmic frontier can be found right on our doorstep. The Galaxy that we inhabit and the galaxies closest to us embody the complete evolutionary history of the Universe up to the present day. Whatever can have happened has left telltale signs in the sundry observable components of our Galaxy and galaxian neighbors. The distributions of mass, age, kinematics, and elemental abundance all give crucial clues to the organizing and transformative processes that have occurred over cosmic time.

In the Milky Way, astronomers have calculated the ages of globular clusters as 12–14 giga-years, thus placing a lower limit on the age of the Universe. The types of stars populating the disk further indicate that a burst of star formation occurred some 12 giga-years ago. If the Milky Way is representative, the formation of massive disks was probably endemic during that early epoch. The presence of low-mass satellite galaxies around the Milky Way demonstrates that close interactions and mergers continue to modify the numbers and properties of galaxies. But the thinness of the Milky Way's disk suggests that a major disruption has not occurred for billions of years.

The disks of most other spiral galaxies are also remarkably thin. Moreover, the velocity fields in the disks are predominantly circular, and hence are regular enough to host spiral density waves and other relatively quiescent dynamics. The galaxies that do show significant disruption are relatively rare. This paucity of seriously perturbed disk galaxies places a constraint on the rate of strong interactions during the last few billion years.

Elliptical galaxies and the bulges of spiral galaxies belie the placid scenario presented by the disks. Swarming rather than rotating, these stellar conflagrations either fell into place a very long time ago or have been whipped into a frenzy from subsequent merger and inflow events. The bulges of spiral galaxies are especially vexing, for they can contain many generations of stars from a wide assortment of gaseous inflow events. Just as all paths lead to Rome, many paths in a galaxy may ultimately lead to its bulge.

The colors in spiral galaxies typically show a gradient of increasing blueness at larger radii. The elemental abundances are seen to *decrease* at larger radii in commensurate fashion. These radial gradients are best explained by an outward-going progression of star formation in the spiral galaxies. The bulges and inner disks were the first to transmute from gas to stars, with subsequent generations rapidly building up the abundances of heavy elements. The outer disks, having evolved the least, display colors characteristic of the most recent star formation, with scant chemical enrichment to show for it.

The radial gradients also argue against wholesale redistributions of stars and gas in spiral galaxies. Too much influx of young blue stars from the outer disk would wipe out the color gradients, while strong inflows of primordial gas would erase the abundance gradients. Therefore, the basic spiral Hubble types that we see today are probably close to those that the galaxies had billions of years ago.

The dwarf galaxies, being the most numerous, help to trace out the underlying distribution of mass in the Universe. If galaxy formation was enhanced in slightly overdense regions of dark matter, then the dwarf galaxies should manifest this "bias." Deep surveys of voids in the distribution of bright galaxies, in fact, show a corresponding deficit of dwarf galaxies, and hence a biasing against the voids.

Disk galaxies with low surface brightness continue to confound. Though gas rich, these systems show a paltry record of star formation over cosmic time. Why they have delayed their metamorphosis from gas to stars may be related to the diffuse state of their gas. If so, they may represent intermediate cases between "normal" luminous galaxies and the invisible blobs of gas that make up the quasar absorption-line systems.

The final frontier lies within ourselves, for we have benefited tremendously from the chemical evolution that has transpired in the cosmos. The protons that make up the hydrogen in our tears were forged during the first few billionths of a second following the Big Bang. The calcium in our bones, the iron in our blood, and the gold that adorns our skins came billions of years later—in the explosions of massive stars that once delineated the spiral arms of our Galaxy. It is in many ways reassuring that this sort of ongoing starburst activity has characterized the most transformative epochs of the galaxian Universe, going all the way back to the first few billion years. We are indeed starstuff, as connected to the Universe as the galaxy that spawned us.

Humanizing the Cosmos

Having pondered the full expanse of cosmic history, we are inevitably drawn to the marvel of our very existence. In the last two centuries, stunning progress has been made in elucidating this tale of emergence. And with each new realization, the tracery of our heritage grows ever more exquisite.

In the mid-nineteenth century, Charles Darwin and Alfred Russel Wallace independently came upon the mechanism of natural selection that governs the survival of new species—from the lowly parameciums to the clever *Homo sapiens.* A century later, Francis Crick and James Watson traced out the double helix molecule of DNA (deoxyribose nucleic acid) that underlies all life on Earth. The concurrent discovery of cosmic rays and terrestrial radioactivity—both byproducts of prior starburst activity—revealed the cosmic potential for mutating the genetic sequence within each DNA molecule and thus driving the evolutionary process. And in the first decade of the twenty-first century, molecular biologists have succeeded in fully mapping the human genome, an accomplishment that promises to revolutionize evolutionary biology.

As we endeavor to link the uncanny nature of our own existence to the unrelenting flow of cosmic history, what new discoveries await us? At the risk of divining the future, we offer some clues in the Epilogue.

EPILOGUE

There remain several big questions that linger on the cosmic frontier without ready solution. These quandaries are so compelling, however, that they are impossible to ignore. So, we end this book by speculating upon two of the major unknowns: the future of the cosmos, and our role as humans in it.

A Cosmic Forecast

What is our Universe coming to? For the past 10 billion years or so, galaxies have dominated the cosmic scene. Within the galaxies, massive stars have come and gone, while stars like the Sun have quietly nestled in for the long term. What are we to expect in the next 10 billion years, or looking much further ahead, the next 1,000 trillion years?

If we restrict our predictions to what we can directly detect, we find rather bleak prospects. Because the visible matter in galaxies is a factor of 100 less than that necessary to gravitationally bind the expanding cosmos, galaxies will continue to recede from one another. Those galaxies that happen to be bound members of rich clusters will be subject to occasional gravitational perturbations by their neighbors. Over time, these perturbations will accumulate so that some of the galaxies will escape the cluster, while others will fall into whatever giant elliptical galaxy is closest to the cluster's center of gravity. We can see similar dynamics already at work—but on much smaller scales—in globular star clusters, where the inner stars are settling toward the center while the outer stars are evaporating. The eventual outlook for each cluster of galaxies will feature central hyper-massive black holes amidst a scattering of intergalactic tramp galaxies.

Within the galaxies themselves, gravitational perturbations by central bars and companion galaxies will ultimately yield similar redistributions of matter. Some stars will lose kinetic energy and fall into the galaxian cores, while others will gain sufficient energy to escape. Once again, but on smaller scales, we see central super-massive black holes occasionally flaring with each stellar plunge into the abyss—all awash in a haze of high-velocity stars.

In the next 10 billion years, stars like the Sun will die off, leaving behind slowly cooling white dwarfs. In the next 100 billion years, no more starbirth will replenish the stellar hoard, the remnant gas in the galaxies having been used up long ago. In the next trillion years, stars with masses a tenth that of the Sun will expire, with the higher-mass white dwarfs having cooled into obscurity. And in the next 1,000 trillion years, even the lowest-mass stars will be dead. Save for the occasionally flaring black hole and merging binary system of stellar remnants, the Universe will be utterly dark. Even the 3-degree cosmic background radiation will have cooled to an equivalent temperature of a millionth of a degree—its characteristic wavelength measured in kilometers, and its surface brightness a trillion-trillion (10^{24}) times fainter than it is today.

If we allow for an overwhelming abundance of dark matter, then our cosmic prognosis becomes less extreme. Strong support for the claim that dark matter constitutes 85–95 percent of the mass in the Universe comes from the kinematics of gas within galaxies and of galaxies within galaxy clusters. These motions require a degree of gravitational binding that greatly exceeds the gravitation associated with the luminous matter alone. Observations of distant galaxies and quasars that have been gravitationally lensed by foreground clusters of galaxies also indicate copious amounts of dark matter within the clusters in order to explain the high degree of lensing (see Plate 29).

In the presence of the ponderous dark matter, star formation will continue to decline, and the stars themselves will die out, but on timescales that depend on the nature of the dark matter. Meanwhile, the inertia of the dark matter will slow the inward-outward scattering dynamic in galaxies and galaxy clusters that would otherwise lead to a hierarchy of black holes surrounded by renegade stellar remnants.

Although the currently inferred amount of dark matter is about a factor of 3 less than that necessary to bind the expanding Universe, the still uncertain nature of the dark matter has interesting consequences. If baryons, such as protons and neutrons, dominate the dark matter, then there is plenty of raw material in the Universe to make new galaxies and stars. All we have to do is wait for the filamentary intergalactic gas to congeal into recognizable galaxies. If photinos, gravitinos, magnetic monopoles, or other exotic forms of cold dark matter prevail, then entirely different constructs may ensue. We can only guess what photino-dominated galaxies might eventually look like. And if neutrinos or other forms of hot dark matter abound, then we might expect the material Universe to end up being organized into colossal structures measuring billions of light-years across and containing a plethora of black holes and stellar remnants.

Finally, if our expanding Universe is entering an accelerated phase driven by some sort of dark energy, we are left speculating on the ultimate fate of the cosmos. Will it expand forever, or will it eventually recompress on itself? Can sister or daughter universes coexist with a universe dominated by dark matter and dark energy, or is our space-time all there is? At this point, we can only imagine the possibilities. Perhaps some day, having more fully explored what we regard today as the cosmic frontier, our descendants will know enough to answer these questions—and to imagine even more wondrous possibilities.

The Conscious Universe

Given our capacity to ponder the past, present, and future, to propagate our findings across generations, and to build from them ever greater scientific paradigms of physical reality, it is tempting to think that our species holds a special place in the Universe—some role of cosmic import. Such anthropocentric thinking has helped to galvanize religious belief for thousands of years. More recently, contemporary scientists have begun to weigh in on the topic.

Some contend that our very existence critically depends on the Universe being the way it is. If the history of cosmic expansion were slightly different, or the constants of gravity and electromagnetism a bit askew, galaxies of stars and planets would not have formed, and the complex chemistry of life would have been nixed. This sort of argument is known as the weak version of the anthropic principle. Other scientists have more boldly claimed that the Universe is the way it is *so that* a species capable of pondering its cosmic origins and fate would emerge. This strong version of the anthropic principle places humanity squarely at the focus of a purposeful Universe—one whose raison d'être is to become fully conscious.

As a scientific theory, the strong version of the anthropic principle is difficult to refute or support. Moreover, it reeks of more than a little bit of hubris. Nonetheless, some cosmologists have allowed that it is worthy of consideration, as they make theoretical universes to compare with the one at hand. The idea that our Universe is but one of myriad space-times spawned by the quantum vacuum has given cosmologists a new way of dealing with the anthropic question. By varying all of the possible physical parameters, they are beginning to find out which kinds of space-times are most amenable to fostering long-lasting sources of reliable energy, such as stars, and the complex chemicals such as DNA necessary for conscious life.

What they have found is that our particular Universe has been exquisitely tuned to bring forth the physical, chemical, and biological evolution that we have come to know. But there are many parameters to jiggle—especially now, with the reemergence of the cosmological constant as a viable counterpoint to gravitational attraction. Depending on the balance of gravity and the antigravity that is embodied in the cosmological constant, cosmologists can concoct a wide variety of universes whose evolutionary scenarios might ultimately lead to fully conscious life.

Back in our observable Universe, evidence has been slowly accruing in favor of widespread chemical complexity. Interstellar clouds in the Milky Way and in other galaxies play host to a witch's brew of organic molecules, while the nearest star-forming regions are sodden with water vapor. Within the Solar System, liquid water dominates the chemistry of Earth, and is strongly suspected to exist just below the surfaces of Mars and the Jovian satellite Europa. The far-flung comets become veritable fountains of water when they pass close to the Sun. And the meteorites collected on Earth include a type of carbonaceous chondrite that is rich in amino acids, the building blocks of basic proteins.

In 1961, Frank Drake of the National Radio Astronomy Observatory estimated the number of stars in our Galaxy, and then winnowed that number by formulating the fraction of stars that could be long-lived hosts of planets, on which water-based complex chemistry could ultimately lead to fully concious and technologically capable lifeforms. By formalizing the various requirements for generating a technologically communicative civilization, the eponymous Drake Equation elucidates a harrowing sequence of (im)pobabilities. Yet the Universe is an awfully big place—beginning with our Milky Way galaxy and extending out to the billions of other galaxies that we can observe. For now, we would be foolish to discount the probability that other cosmically aware civilizations exist in the Milky Way, or elsewhere in the Universe.

Then, what of our special role of cosmic import? Do we share this role with others, or do we alone embody the conscious Universe? Either way, we appear destined to become ever more "cosmic" in our outlook. Like our forebears on the African savannah, we have no way of initially knowing what or who lies beyond the immediate horizon. It is through intrepid seeking that our species has come this far. And it will be through our continuing exploration of the Solar System, Milky Way, and Universe, that we will fulfill whatever cosmic role may await us. The adventure has just begun!

SELECTED
READINGS

General Astronomy

Audouze, J., and G. Israel, ed. *The Cambridge Atlas of Astronomy,* 3rd ed. Cambridge, Eng.: Cambridge University Press, 1994. The most comprehensive and colorful compendium of general astronomy currently available.

Burbidge, G., A. Sandage, and F. H. Shu, ed. *Annual Reviews of Astronomy and Astrophysics,* Palo Alto, Calif.: Annual Reviews, Inc. A yearly series of scholarly articles, each volume containing 10–20 reviews of particular research topics in astronomy and astrophysics, written by respected practitioners in their respective fields.

Burnham, R. Jr. *Burnham's Celestial Handbook,* 3 vols. New York: Dover Publications, 1978. This three-volume magnum opus is organized according to constellation. All sorts of stars, clusters, nebulae, and galaxies in the direction of a particular constellation are described along with relevant mythologies, poems, and astronomical histories. A recognized masterpiece.

Galilei, G. *Sidereus Nuncius.* Venice, 1610. Trans. by A. van Helden as *The Starry Messenger.* Chicago: University of Chicago Press, 1989. Galileo describes his pioneering telescopic observations of the Moon, Venus, Jupiter, the Sun, and the Milky Way.

Krupp, E. C. *Beyond the Blue Horizon: Myths and Legends of the Sun, Moon, Stars, and Planets.* Oxford, Eng.: Oxford University Press, 1991. A treasure trove of myths associated with the day and night sky.

Lang, K. R. *Astrophysical Formulae,* 3rd ed., 2 vols. Astronomy and Astrophysics Library. New York: Springer-Verlag, 1999. As the title suggests, this two-volume edition is packed with tabulations and formulas relevant to astronomy—from nuclear and atomic processes to solar and planetary physics to galactic and cosmological astrophysics. Many of the formulations are presented in an expository manner supported by an abundance of pertinent references.

Malin, D. *A View of the Universe.* Cambridge, Eng.: Cambridge University Press, 1994. A stunningly colorful book of celestial imagery with accompanying commentary by the renowned Australian astrophotographer David Malin.

Norton, A. P. *Norton's 2000.0 Star Atlas and Reference Handbook,* 18th ed. Ed. I. Ridpath. Harlow, Eng.: Longman, 1998. A timeless classic; it includes sky maps plus a full introduction to the celestial objects contained therein. Also includes mappings of the sky in Galactic coordinates.

Raymo, C. *365 Starry Nights.* Englewood Cliffs, N.J.: Prentice Hall, 1982. A delightful book that presents the night sky month by month, day by day. Full of hand-drawn illustrations of the sky, constellations, and astronomical wonders. Written with flare by this oft-featured science columnist for the *Boston Globe.*

Seeds, M. A. *Horizons: Exploring the Universe,* 7th ed. Pacific Grove, Calif.: Brooks/ Cole, 2002. One of many full-color introductory textbooks on astronomy. Besides providing end-of-chapter questions and problems, an interactive website, and multiple teacher supplements, this particular textbook includes a CD-ROM containing "The Sky™ Student Edition" desktop planetarium software package.

Shu, F. *The Physical Universe.* Mill Valley, Calif.: University Science Books, 1982. A popular physics-based textbook at the intermediate undergraduate level, covering material similar to that in Zeilik and Gregory (see below) but in a more expository way. Although the book is becoming somewhat dated, the content holds up surprisingly well.

Tirion, W., and R. Sinnott. *Sky Atlas 2000.0,* 2nd ed. Cambridge, Mass.: Sky Publishing Corporation, 1998. An oversized atlas of the sky including stars, star clusters, nebulae, and galaxies. A great book for getting your bearings straight.

Zeilik, M., and S. Gregory. *Introductory Astronomy and Astrophysics,* 4th ed. Orlando: Harcourt Brace, 1997. An upper-division undergraduate textbook that assumes prior knowledge of physics and calculus, but no formal background in astronomy.

Galaxies

Arp, H. C. *Atlas of Peculiar Galaxies.* Pasadena: California Institute of Technology, 1966. A sobering reminder that not all galaxies are regularly shaped spirals or ellipticals. The gray-scale images reveal a dizzying variety of morphologies and interactive behavior.

Arp, H. C., and B. Madore. *Catalogue of Southern Peculiar Galaxies and Associations,* Cambridge, Eng.: Cambridge University Press, 1987. The Southern Hemispheric counterpart to Arp's *Atlas of Peculiar Galaxies.*

Baade, W. *Evolution of Stars and Galaxies.* Ed. C. Payne-Gaposhkin. Cambridge, Mass.: MIT Press, 1975. In 1958, Walter Baade gave a series of lectures at Harvard College Observatory. The resulting tape transcriptions were extensively reworked by Cecilia Payne-Gaposhkin into publishable form. Therein, the old and young stars that populate our Milky Way galaxy and other galaxies of the Local Group are addressed in terms of evolutionary scenarios that were first pioneered by Baade.

Binney, J., and M. Merrifield. *Galactic Astronomy.* Princeton: Princeton University Press, 1998. A thorough phonomenological description of the contents and kinematics of the Milky Way galaxy and other nearby galaxies. Written at the upper-division undergraduate and beginning graduate student level, but useful for professional astronomers as well.

Bok, B. and P. F. Bok. *The Milky Way,* 5th ed. Cambridge, Mass.: Harvard University Press, 1981. A descriptive account of our home galaxy, by Bart and Priscilla Bok, who together served for many decades as ambassadors of astronomical pursuit and good will.

Combes, F., P. Boissé, A. Mazure, and A. Blanchard. *Galaxies and Cosmology,* 2nd ed. New York: Springer Verlag, 2002. A quantitative textbook by European authors for students at the advanced undergraduate and graduate level. It blends observational results and theoretical modeling, and includes references and exercises at the end of each chapter.

de Vaucouleurs, G., A. de Vaucouleurs, and H. G. Corwin. *Second Reference Catalogue of Bright Galaxies,* 2 vols. New York: Springer-Verlag, 1991. This two-volume catalogue lists the basic properties of 23,024 galaxies down to a *B-band* magnitude of 15.5, including positions, morphological types (on the de Vaucouleurs system), angular sizes, optical and infrared magnitudes, colors, Galactic and internal extinctions, neutral hydrogen emission-line intensities and Doppler widths, and radial velocities. The most comprehensive tabulation of the nearby galaxian Universe.

Elmegreen, D. M. *Galaxies and Galactic Structure.* Upper Saddle River, N.J.: Prentice Hall, 1998. A textbook for undergraduate students with introductory backgrounds in physics, astronomy, and calculus. Includes an abundance of exercises, unsolved problems, useful websites, and other references at the end of each chapter.

Ferris, T. *Galaxies* New York: Stewart, Tabori and Chang, 1982. A lavishly illustrated introduction to the Milky Way, other galaxies, and cosmology, written with panache by an award-winning science writer.

Herschel, W. *On the Construction of the Heavens* (1785) In *The Scientific Papers of Sir William Herschel* London: The Royal Society and the Royal Astronomical Society, 1912. The Herschels' star-gauging efforts and their resulting map of the Milky Way as viewed from outside are documented herein.

Hodge, P. W. *Galaxies.* Cambridge, Mass.: Harvard University Press, 1986. The fourth book on galaxies in the series Harvard Books on Astronomy.

Hubble, E. *The Realm of the Nebulae.* New Haven: Yale University Press, 1936. Published eleven years after the discovery of galaxies as autonomous systems and seven years after the discovery of universal expansion, this popularization by the champion of these epochal discoveries paved the way for subsequent, more eloquent books on the topic.

Kennicutt, R. C. Jr., F. Schweizer, and J. E. Barnes. *Galaxies: Interactions and Induced Star Formation.* Saas-Fee Advanced Course 26, Lecture Notes 1996, Swiss Society for Astrophysics and Astronomy. Berlin: Springer-Verlag, 1998. Three full expositions on induced starbirth and starburst activity, observational evidence for interactions and mergers, and the dynamics of galaxy interactions, by experts in their respective fields.

Sandage, A. *The Hubble Atlas of Galaxies.* Washington, D.C.: Carnegie Institution of Washington, 1961. The essential picture-book of galaxies, originally inspired by Hubble himself and meticulously crafted with high-quality black-and-white images and cogent commentary by Hubble's protégé, Allan Sandage.

Sandage, A., and J. Bedke. *Atlas of Galaxies Useful for Measuring the Cosmological Distance Scale.* Washington, D.C.: NASA, 1988. An oversized coffee-table of a book showing gray-scale images of nearby galaxies with details sufficient to resolve individual bright stars in many of them.

———*The Carnegie Atlas of Galaxies,* 2 vols. Washington, D.C.: Carnegie Institution of Washington and the Flintlock Foundation, 1994. The *ne plus ultra* of galaxy atlases, containing gray-scale images of 1,168 galaxies listed in the *Shapley-Ames Catalog,* segregated according to morphological type.

Sandage, A., and W. Tammann. *A Revised Shapley-Ames Catalog of Bright Galaxies,* Washington, DC: Carnegie Institution of Washington, Publication 635, 1981. In this combination catalog and atlas, the authors classify and tabulate the properties of 1,246 galaxies, illustrating each morphological type with greyscale prints of representative galaxies.

Shapley, H. *Galaxies.* Cambridge, Mass.: Harvard University Press, 1943. The first book on galaxies in the series Harvard Books on Astronomy.

Sinnott, R. W. *NGC 2000.0, The Complete New General Catalogue and Index Catalogue of Nebulae and Star Clusters by J. L. E. Dreyer,* Cambridge, Eng.: Cambridge University Press, 1988. An updated edition of Dreyer's 1888 catalogue, featuring all of the NGC and IC listings.

Sparke, L. S., and J. S. Gallagher. *Galaxies in the Universe.* Cambridge, Eng.: Cambridge University Press, 2000. A complete textbook on galaxian astronomy written at the upper-division undergraduate level. Concepts, graphics, derivations, references, and problems are woven together in each of the eight chapters, which together span the full range of galactic and extragalactic astronomy.

Tully, R. B., and J. R. Fisher. *Nearby Galaxies Atlas.* Cambridge, Eng.: Cambridge University Press, 1987. An oversized atlas presenting the positions of 2,367 galaxies as they lie on the sky, and stunningly graphic visualizations of galaxy groups, clusters, and superclusters in three dimensions. Accompanying tabular data are presented in the *Nearby Galaxies Catalog,* by R. B. Tully. Cambridge, Eng.: Cambridge University Press, 1988.

van den Bergh, S. *Galaxy Morphology and Classification.* Cambridge, Eng.: Cambridge University Press, 1998. A 111-page monograph on the various systems of classifying galaxies according to their shapes. It includes chapters on the Hubble, de Vaucouleurs, Elmegreen, van den Bergh, and Morgan systems, along with chapters on the different kinds of galaxies in the local Universe, the evolution of galaxy morphology over cosmic time, and computerized classification schemes.

Wray, J. D. *The Color Atlas of Galaxies.* Cambridge, Eng.: Cambridge University Press, 1988. Contains full-color images of nearby galaxies spanning the entire sequence of Hubble types, with commentary on the stellar, nebular, and dust content in these diverse realms.

Cosmology

Adams, F., and G. Laughlin. *The Five Ages of the Universe: Inside the Physics of Eternity.* New York: The Free Press/Simon & Schuster, 1999. A chronology of the Universe spanning more than 100 orders of magnitude in time. Considers the primordial, stelliferous, degenerate, black hole, and dark eras as foreseen from the present epoch.

Gamow, G. *The Creation of the Universe.* New York: Viking, 1952. The first popularization of what was to become known as the Hot Big Bang theory, written with characteristic verve by one of the greatest story-tellers in science.

Greenstein, G. *The Symbiotic Universe: Life and Mind in the Cosmos.* New York: William Morrow and Co., 1988. Ruminations on the emergence of life in the Universe, the anthropic principle, quantum physics, and consciousness.

Harrison, E. R. *Cosmology: The Science of the Universe,* 2nd ed. Cambridge, Eng.: Cambridge University Press, 2000. A clearly written textbook on the origin, geometry, and evolution of the expanding Universe, aimed at intermediate-level undergraduate students having some background in physics and calculus. Recently updated.

———*Darkness at Night: a Riddle of the Universe.* Cambridge, Mass.: Harvard University Press, 1987. An erudite exposition on Olbers' Paradox and its cosmological significance.

Hawking, S. *A Brief History of Time: From the Big Bang to Black Holes,* 10th anniversary ed. New York: Bantam Books, 1998. Tracks both the history of the cosmos and the quest to understand our cosmic origins. Written with straightforward clarity and good-natured humor by a legendary physicist.

Hogan, C. J. *The Little Book of the Big Bang: A Cosmic Primer.* New York: Copernicus, 1998. A gem of a book, complete with all the essentials for understanding cosmic expansion, the geometry of space-time, inflationary models, and recent controversies in quantum cosmology.

Hoyle, F., G. Burbidge, and J. Narlikar. *A Different Approach to Cosmology.* Cambridge, Eng.: Cambridge University Press, 2000. As the title suggests, this provocative book addresses universal expansion and other cosmological findings from a different perspective. After laying out various difficulties with the standard Hot Big Bang model, the authors propose a modified version of the steady-state theory as the more viable alternative. Written at the graduate/professional level.

Narlikar, J. V., and F. Hoyle. *An Introduction to Cosmology,* 3rd ed. Cambridge, Eng.: Cambridge University Press, 2002. A full quantititative introduction to the field of cosmology written at a level suitable for advanced undergraduate and graduate students.

Overbye, D. *Lonely Hearts of the Cosmos.* New York: HarperCollins, 1991. This engaging exposé recounts the human stories behind the emerging science of cosmology—from Hubble to Hawking.

Poe, E. A. *Eureka* (1848) Reprinted in the *The Science Fiction of Edgar Allan Poe,* ed. H. Beaver. Harmondsworth, Middlesex, Eng.: Penguin Books, 1976. This essay features Poe's thoughts on why the night sky is dark (Olbers' Paradox) and speculations on the ultimate fate of the Universe.

Silk, J. *The Big Bang,* 3rd ed. New York: W. H. Freeman, 2001. An excellent introduction to cosmology (the origin and overall evolution of the Universe) and cosmogony (the formation of galaxies, stars, and planets), written in a descriptive yet thorough manner. Mathematical derivations and a glossary are included at the end as a supplement to the text.

———*A Short History of the Universe.* New York: W. H. Freeman, 1994. This fifty-third book in the Scientific American Library series is written, illustrated, and produced with the polish familiar to readers of *Scientific American.* The origin and early evolution of the Universe are covered from an active cosmologist's viewpoint.

Weinberg, S. *The First Three Minutes: A Modern View of the Origin of the Universe,* 2nd ed. New York: Perseus Books LLC, 1993. A descriptive accounting of the Hot Big Bang back to the first one-hundredth second, emphasizing the physics of elementary particles at these early epochs and the correspondingly high temperatures and densities. An afterword describing progress in cosmology since the first edition came out in 1977 is included. A glossary, mathematical supplement, and annotated list of references are also included.

PERIODICALS
AND WEBSITES

Periodicals

American Scientist, the magazine of Sigma Xi, the Scientific Research Society, Research Triangle Park, N.C.: Sigma Xi.
http://www.americanscientist.org
Astronomy magazine, Waukesha, Wis.: Kalmbach Publishing Co.
http://www.astronomy.com/
Mercury magazine, San Francisco: Astronomical Society of the Pacific.
http://www.aspsky.org
Nature magazine, Washington, D.C.
http://www.nature.com
Science magazine, Washington, D.C.: American Association for the Advancement of Science.
www.sciencemag.org
Scientific American magazine, New York: Scientific American.
http://www.sciam.com
Sky & Telescope magazine, Cambridge, Mass.: Sky Publishing Corporation.
http://www.skypub.com

Websites

Anglo-Australian Observatory (AAO)
http://www.aao.gov.au
Astronomy Picture of the Day (APOD)
http://www.antwrp.gsfc.nasa.gov

Cosmology at the National Center for Supercomputing Applications
  http://archive.ncsa.uiuc.edu/Apps/Cosmo
European Southern Observatory
  http://www.eso.org
Galaxies and the Cosmic Frontier
  http://cosmos.phy.tufts.edu/cosmicfrontier
Harvard-Smithsonian Center for Astrophysics
  http://cfa-www.harvard.edu
Harvard University Press
  http://hup.harvard.edu
International Dark-Sky Association
  http://www.darksky.org
W. M. Keck Observatory
  http://www2.keck.hawaii.edu
Kitt Peak Advanced Observing Program
  http://www.noao.edu/outreach/aop
National Aeronautics and Space Administration (NASA)
  http://www.nasa.gov
NASA Astronomical Data System (ADS) literature search engine
  http://adsabs.harvard.edu/abstract_service.html
NASA Space Science
  http://spacescience.nasa.gov
NASA/IPAC Extragalactic Database (NED)
  http://ned.ipac.caltech.edu.
National Optical Astronomy Observatories (NOAO)
  http://www.noao.edu.
National Radio Astronomy Observatory (NRAO)
  http://www.nrao.edu
SkyView Virtual Observatory
  http://skyview.gsfc.nasa.gov
Space Telescope Science Institute (STScI)
  http://www.stsci.edu
SPACE.com multimedia e-magazine
  http://www.space.com
Students for the Exploration and Development of Space
  http://www.seds.org
Subaru Telescope: National Astronomical Observatory of Japan
  http://www.naoj.org

GLOSSARY

absolute luminosity The radiant power of an object measured as the amount of energy emitted per unit of time. Commonly used units are ergs per second, joules per second (Watts), and solar luminosities, where 1 solar luminosity equals 3.84×10^{33} ergs/second or 3.84×10^{26} Watts.

absolute magnitude The luminosity of an object measured in the logarithmically compressed units of magnitudes and defined as the apparent magnitude the source would have if located at a distance of 10 parsecs (32.6 light-years). The Sun has an absolute visual magnitude of $+4.83$.

absorption spectrum A spectrum of an object in which cool gases have absorbed light of certain wavelengths, producing gaps in the brightness at these particular wavelengths.

active galactic nuclei (AGNs) Nuclei found in galaxies that exhibit unusually energetic behavior in their centers.

amorphous galaxy A galaxy type that has no discernible regularity of structure.

Angstrom A unit of length used in spectroscopy, equal to 10^{-10} meters.

antimatter A subatomic form of matter that is of opposite charge to that of ordinary matter. Examples of antimatter include antiprotons, antielectrons (positrons), antineutrinos, and antiquarks.

apparent magnitude The brightness of an object as seen from the Earth, measured in units of magnitudes. The brightest stars in the sky are of first magnitude or lower; the faintest naked-eye stars are of magnitude 6. The most powerful telescopes can reach magnitudes as faint as 30.

arcsecond (arcsec)   A second of arc, a measure of angular extent. One arcsecond equals 1/3,600 of a degree.

association   A group of stars, usually young, massive stars, that share a common origin. Also called OB association.

atom   The basic building block of ordinary gaseous, liquid, and solid matter. It consists of a central nucleus containing protons and neutrons, with a surrounding cloud of electrons. A neutral atom has equal numbers of oppositely charged protons and electrons.

atomic number   The number of protons and corresponding positive charge in the nucleus of an atom. Each element is specified by its atomic number.

atomic weight   The combined number of protons and neutrons in the nucleus of an atom. Different isotopes of the same element are differentiated by their atomic weight.

Balmer lines   The spectral lines of atomic hydrogen that appear in the visible part of the spectrum, either in absorption or in emission. They are produced by electronic transitions up from or down to the first energy level above the groundstate.

barred spiral   A galaxy with a distinctly linear central feature in its overall structure.

baryons   A class of heavy subatomic particles among the hadron family of elementary particles. Baryons include protons and neutrons, and are thought to consist of quark triplets.

Big Bang   The hypothesized event some 15 billion years ago that exploded from a state of extremely high density and temperature to the Universe that we now know.

binary   A double star or galaxy whose two components revolve around each other's mutual center of mass.

black body   An idealized object that absorbs all radiation that falls on it and reemits it at a rate that maintains a state of thermal equilibrium. The spectrum of black-body radiation is completely specified by the body's temperature according to known laws.

black hole   An object that is so massive and compact that neither light nor anything else can escape from its gravity.

blazar   A quasar-like object that has a continuous spectrum without line emission.

boson   A quantum particle of integer spin angular momentum. Unlike fermions, bosons can share the same quantum state and so do not obey the Pauli exclusion principle. Bosons include photons (conveyers of the electromagnetic force), W and Z bosons (conveyers of the weak nuclear force), and mesons.

Brackett lines   The lines of atomic hydrogen that appear in the infrared part of the spectrum, either in absorption or in emission. They result from electronic transitions up from or down to the third energy level above the groundstate.

Bremsstrahlung   Radio wavelength continuum emission originating from ionized hydrogen, as free electrons accelerate past the hydrogen nuclei (protons).

bulge   In a galaxy, the area around the nucleus of a spiral galaxy where stars are arranged in a less flattened volume than in the surrounding disklike area.

cD galaxy   A large, diffuse galaxy, often the brightest in a cluster of galaxies.

Cepheid variable   A type of pulsating supergiant star with a period of variability of between about 1 and 100 days. The well-calibrated relation between a Cepheid variable's period and its luminosity is used in establishing distances to nearby galaxies. The term was named for the prototype variable star, Delta Cephei.

Chandrasekhar limit   The upper mass limit of a white dwarf, above which the stellar remnant will gravitationally collapse to a neutron star or explode. It was derived by S. Chandrasekhar in 1931 to be 1.4 solar masses.

cluster   A physical group of stars that were born together or of galaxies that share a common origin.

cold dark matter (CDM)   Undetected material with significant mass and relatively low characteristic velocities.

color-magnitude diagram (CMD)   A graphical plotting of colors and magnitudes of stars in a cluster or galaxy from which various characteristics of the stellar population can be inferred.

continuous spectrum   The light emitted by an object at all wavelengths and without gaps caused by absorption at particular wavelengths. The rainbow is an example of a continuous spectrum.

corona   The outermost region of an astronomical object, such as the Sun or the Milky Way.

corotation radius   The distance from the center of a spiral galaxy where the stars and gas rotate at the same rate as the spiral density wave.

cosmic microwave background radiation (CMBR)   The radiation left over from the epoch when the cooling Universe changed from an ionized plasma to a neutral atomic state, approximately 300,000 years after the Big Bang.

cosmic rays   High-energy charged particles that originate beyond the Earth and are detected when they penetrate the Earth's atmosphere.

cosmological constant   A constant in the equation of state of general relativity, introduced by Albert Einstein, who later retracted it. It is now thought to be nonzero.

cosmology   The study of the large-scale Universe.

dark energy   An all-pervasive energy in the Universe with no relation to any known radiating or absorbing objects. It could be driving the accelerating expansion of the Universe that has been measured recently.

dark matter   Ponderous material of an unknown kind that is manifested in the motions of stars in galaxies, of galaxies in clusters of galaxies, and of light rays traversing these massive objects.

dark nebula   A cloud of gas and dust that obscures the stars or galaxies behind it.

degeneracy   An unusual state of matter where the Pauli exclusion principle applies, so that the pressure depends on density alone, with no sensitivity to temperature. The electrons in a white dwarf star are degenerate, exerting sufficient pressure to support the star against gravitational collapse.

density wave   A periodic dynamical disturbance in the disk of a spiral galaxy in the form of two or more spiral wavefronts of enhanced density and gravity that can self-propagate around the disk and so generate the grand-design spiral structure that is observed in some spiral galaxies.

deuterium   A heavy isotope of hydrogen, H^2, where the nucleus contains one proton and one neutron.

differential rotation The property of a rotating body whose rotation period is not the same at all radii, as opposed to solid-body rotation.

diffuse nebula A cloud of gas and/or dust that is illuminated by starlight.

disk In galaxies, the flat component that encompasses the spiral arms and other young, gas-rich material.

Doppler shift The change in detected wavelength of light or sound caused by the motion of the source toward or away from the observer.

dynamics The gravitational interactions within a body or between bodies in motion.

eccentricity A measure of how elongated an ellipse is, defined as the ratio of the distance between the foci to the length of the major axis of the ellipse. A circle has an ellipticity of zero; a straight line is equivalent to an ellipse with an ellipticity of one.

electromagnetic radiation All types of lightlike radiation involving oscillating electric and magnetic fields, ranging from short wavelengths and high frequencies (gamma rays) to long wavelengths and low frequencies (radio waves).

electron A negatively-charged particle belonging to the lepton family of elementary particles, with a mass that is 1/1,836 that of the proton. Electrons either are bound to atoms or are unbound components of a plasma.

electron volt The amount of energy acquired by an electron when accelerated through a voltage difference of one volt.

elliptical galaxy A type of galaxy that has a regular elliptical appearance and contains mostly very old stars yellow-red in color.

emission line The bright emission of a particular color (wavelength) in the spectrum of an incandescent or fluorescent gas.

emission nebula A gaseous cloud that emits the emission lines of its constituent gases.

erg A small metric unit of energy in the centimeter-gram-second (CGS) system of units, amounting to the work required to lift 1 gram of mass 1 centimeter above the Earth's surface.

Euclidean Referring to space, a geometry that conforms to Euclid's axioms, so-called flat space.

event horizon   The critical surface surrounding a black hole where the velocity of escape equals the velocity of light.

exponential   Governed by an equation in which a variable is in the exponent (or power) of some constant, such as 10^x or e^x, where x is the variable. Characterized by strong change, as in exponential growth or exponential decrease.

extinction   The reduction of the received intensity of light from a source caused by intervening material, such as dust.

extragalactic   Beyond the realm of our local Galaxy.

fermion   A quantum particle of half-integer spin angular momentum. Fermions obey the Pauli exclusion principle, in that they cannot share the same quantum state. Fermions include baryons (protons and neutrons) and leptons (electrons and neutrinos).

flocculent spirals   Spiral galaxies whose spiral structure is fragmentary and indistinct, reminiscent of sheep's wool.

flux   A measure of received brightness in units of ergs per second per square centimeter at the detector, or equivalently Watts per square meter. The Sun's radiant flux on Earth is 1,380 Watts per square meter. The brightest naked-eye stars have fluxes that are 10 billion times smaller.

forbidden lines   Spectral emission lines that result from electronic transitions that are very improbable but nevertheless occur in the rarefied environment of interstellar space.

force   An interaction resulting in attraction or repulsion. The four known forces are the strong force between quarks in nuclei, the weak force between quarks and leptons, the electromagnetic force between any charged particles, and the gravitational force between any masses.

frequency   The number of vibrations per unit of time, as in cycles/second.

galactic cluster   A small, relatively young (open) cluster of stars in our local galaxy.

galaxy   A large collection of stars and other material components that is gravitationally bound, mostly autonomous with respect to other objects, and usually separated from its neighbors by millions of light-years.

gamma rays   Very short-wavelength, high-frequency, electromagnetic radiation. Gamma-ray photons are of very high energy.

general relativity   The theory of gravitation developed by Albert Einstein between 1906 and 1916, in which gravity is manifested by the curvature of space-time and the consequent trajectories of matter and light.

giant branch   The portion of the color-magnitude diagram where stars reside after their main-sequence lifetimes. The stars are typically brighter and often redder than they were on the main sequence.

gigahertz (Ghz)   A unit of frequency equal to 10^9 Hertz (cycles/second).

giga-years (Gyr)   A unit of time equal to 10^9 years.

globular cluster   A usually massive, compact group of stars that is gravitationally stable. The globular clusters of the Milky Way are very old.

gluon   The fundamental particle that conveys the strong force between nuclear particles.

grand-design spiral   A spiral galaxy with a prominent two-armed spiral structure.

grand unified theories (GUTs)   Attempts to unify the detailed understanding of matter and energy at the subatomic level.

gravitational instability   The property of an assemblage that describes its susceptibility to collapse under its own self-gravity.

gravitational lens   A massive object whose gravitational field bends the light from a background source.

gravitational radiation   Propagating oscillation of the gravitational field, analogous to the electric and magnetic field oscillations that constitute electromagnetic radiation (light). Gravitational radiation is generated by gravitating matter that is accelerating, as in a close binary star system, a rapidly rotating neutron star, or an imploding stellar core just before a supernova explosion.

graviton   The hypothesized quantum particle that carries the gravitational force.

groundstate   The lowest allowed energy level.

H I   Neutral hydrogen gas.

H II   Ionized hydrogen gas.

H II region   A gas cloud that contains mostly hydrogen that is ionized.

hadrons A class of subatomic particles that are thought to be made of quarks and to be subject to strong interactions involving the exchange of gluons. Hadrons include baryons such as protons and neutrons, various kinds of mesons, and their antimatter counterparts.

halo An outer envelope or constituent, especially of a galaxy.

Heisenberg uncertainty principle The property of subatomic matter that limits simultaneous measurement of a particle's position and momentum, or energy and timeframe, to a minimum but finite level of uncertainty.

Hertz A unit of wave frequency equal to one cycle per second.

horizontal branch A portion of a color-magnitude diagram where very old stars are found at absolute magnitudes of about 0.7, after having been on the red-giant branch.

hot dark matter (HDM) Undetected material with significant mass and relatively high characteristic velocities.

Hubble constant The constant of proportionality between the distances and the radial (recession) velocities of galaxies.

Hubble law The law of proportionality between the distances and the radial (recession) velocities of galaxies.

Hubble time The age of the expanding Universe calculated from the Hubble law under the assumption that the expansion is neither accelerating or decelerating.

hypernova An anomalously bright supernova.

***Index Catalogue* (IC)** A supplement to the *New General Catalogue of Nebulae and Clusters of Stars* (NGC).

inclination The angle between the plane of an object, such as the disk of a spiral galaxy, and the imaginary plane of the sky. A face-on configuration has zero inclination; an edge-on configuration has an inclination of 90 degrees.

inflation In cosmology, a brief period when the early Universe experienced a radically accelerated rate of expansion.

infrared Light of wavelength just longer than that of visible radiation. The infrared band spans wavelengths of 1 micron (10^{-6} meters) to several hundred microns.

intensity   A measure of received brightness (flux) per unit angular area on the sky (as in ergs per second per square centimeter per square arcsecond), or, equivalently, luminosity per unit area at the source (as in solar luminosities per square parsec).

interstellar medium   The gas and dust that exists between the stars.

ion   An atom that has lost or gained one or more electrons and is thus electrically charged.

ionization   The process of subtracting electrons from or adding electrons to previously neutral atoms, thereby changing a gas of neutral atoms to one made up of charged ions.

irregular galaxy   A type of galaxy with little or no symmetry or spiral structure.

isochrones   Lines of equal age in a color-magnitude diagram.

isophotes   Lines (contours) of constant brightness, as on a map of a galaxy.

isothermal   Having equal temperature, as in a gas of atoms, or having similar average speed, as in a galaxy of stars.

isotope   A variation of an atom, or element, of fixed atomic number that depends on the total number of protons and neutrons (the atomic weight). For example, the isotopes of uranium-238 and uranium-235 differ by their respective numbers of neutrons, the number of protons being 92.

isotropic   Having properties that do not depend on direction.

Jeans length   The characteristic size of a perturbation that is just large enough to be gravitationally unstable.

Jeans mass   The mass of a perturbation of a given size that is just large enough to be gravitationally unstable.

Keplerian   Pertaining to orbits that obey Kepler's laws—having elliptical trajectories and average velocities that decrease as the square root of the ellipse's semi-major axis.

kiloelectron volts   One thousand electron volts, a measure of energy.

kiloparsec (kpc)   One thousand parsecs, amounting to 3,260 light-years.

kinematics   The motions of objects, as in the case of the stars in a galaxy.

leptons   A class of light subatomic particles that includes electrons, neutrinos, and their antimatter counterparts. Leptons are subject to weak, electromagnetic, and gravitational interactions but not to strong interactions.

light curve　A graphical plot of the changing brightness of a source over time.

light-day　The distance light travels in an Earth day (2.6×10^{10} kilometers).

light-year　The distance that light travels in one Earth year (9.46×10^{12} kilometers).

Local Group　The family of about 40 galaxies to which the Milky Way galaxy belongs.

logarithm　The power of 10 that generates a particular number. For example the number 1,000 can be stated as 10^3, its logarithm being 3. The fraction 1/1,000 can be stated as 10^{-3}, its logarithm being -3.

lookback time　The time between the present observation of a distant source and the time when the light from the source was emitted. The lookback time for the Sun is about 8 minutes.

low-ionization nuclear emission region (LINER) galaxies　Galaxies with some nuclear emission, but with relatively low levels of activity in their centers.

luminosity　The intrinsic power of a radiating source, typically measured in units of ergs per second, Watts, or solar luminosities.

luminosity class　An estimate of the luminosity of a star or galaxy based on certain observed criteria that divide these objects into recognizable classes.

luminosity function　A curve or table that tells how many stars or galaxies there are at each absolute luminosity.

Lyman-alpha forest　A complex of Lyman-alpha absorption lines of hydrogen seen in the spectra of some quasars and caused by intervening clouds of hydrogen gas.

Lyman limit　The ultraviolet wavelength of 912 Angstroms, at which the Lyman series of hydrogen lines terminates, corresponding to the minimum energy required to liberate hydrogen's sole electron from the groundstate to an unbound state.

Lyman lines　The spectral lines of atomic hydrogen that appear in the ultraviolet part of the spectrum, either in absorption or in emission. The result from electronic transitions up from or down to the groundstate.

Magellanic Clouds   Two rather small, irregular galaxies that are close companions of the Milky Way galaxy.

magnitude   A measure of the brightness of an astronomical object that uses a logarithmic scale. An increase of 1 magnitude corresponds to a diminution in brightness by a factor of 2.5, and an increase of 5 magnitudes corresponds to a dimming by a factor of $(2.5)^5$, or 100.

main sequence   The diagonal band in a color-magnitude or temperature-luminosity diagram where most stars are found. It represents the locus of temperature and luminosity that characterizes stars of different mass during the most stable and long-lasting part of their lives.

mass   The measure of the amount of matter in an object as determined by its gravitational field or by its response to a force.

massive compact halo objects (MACHOs)   Conjectured to be the dark matter that is concentrated in galaxies and galaxy clusters.

mass-to-light ratio   The ratio of the mass of an object to its luminosity, usually expressed in solar units.

matter-energy   An inclusive term that relates ordinary and exotic forms of matter, electromagnetic radiation, and force fields according to Albert Einstein's mass-energy equation, $E = mc^2$, where E is energy, m is mass, and c is the speed of light.

megaparsec (Mpc)   A million parsecs, amounting to 3,260,000 light-years.

mega-year (Myr)   A unit of time equal to 1 million (10^6) years.

metal   In the parlance of astronomers, any element heavier than helium.

metal-poor   Lacking in heavy elements compared to the Sun.

metal-rich   Similar to the Sun in terms of heavy element abundance.

micron   A measure of length equal to one-millionth of a meter.

micrometer   A measure of length equal to one-millionth of a meter.

microwave   Short-wavelength radio waves with wavelengths between about 1 millimeter to 300 millimeters (30 centimeters).

molecule   An electrically bound assemblage of atoms.

molecular cloud   A cool, dense cloud of gas and dust in which molecules are a conspicuous component.

nebula   A diffuse cloud, usually of gas or dust or both. Historically, nebulae included any extended fuzzy-looking object, including what we today recognize as galaxies.

neutrino   A very small subatomic particle belonging to the lepton family, with no charge, little or no mass, and poor interactivity involving the weak force.

neutron   A subatomic particle belonging to the hadron family with no charge and with a mass slightly more than that of a proton. Outside of the atomic nucleus, the neutron disintegrates into a proton, electron, and neutrino in about 11 minutes.

neutron star   A very dense and compact stellar remnant following the collapse of a massive star's core, or perhaps of a white dwarf remnant that has been accreting mass from a nearby companion star. The collapsed core is comprised primarily of neutrons and has a density similar to that of atomic nuclei. The lower mass limit of a neutron star is 1.4 solar masses, and the upper mass limit is about 3 solar masses—above which the object would collapse to a black hole.

New General Catalogue of Nebulae and Clusters of Stars (NGC)   A catalogue compiled near the end of the nineteenth century.

nova   A star that suddenly increases in brightness owing to the exchange of gases from a more massive star to a small, dense, hot companion star, resulting in a runaway thermonuclear reaction on the dense star's surface.

nucleosynthesis   The process in which larger atoms are built up from smaller ones through termonuclear fusion.

nucleus   The central regions of an object, such as a galaxy.

OB association   A group of very young and hot stars that formed together but that are not necessarily gravitationally bound.

opacity   The ability of a material to impede the passage of light.

open cluster   A group of stars, usually smaller, less condensed, and younger than a globular cluster.

optical depth   The degree to which light has been absorbed by a given amount of material.

parallax   The apparent annual back-and-forth motion of a nearby star with respect to background stars in the sky, resulting from the Earth's orbital motion around the Sun.

parsec   A unit of distance equal to that of a star that exhibits a heliocentric parallax of 1 second of arc. One parsec equals 3.26 light-years, or 3.086×10^{13} kilometers.

Pauli exclusion principle   A property of subatomic matter that excludes two particles of the same type from occupying the exact same quantum state. It is obeyed by fermions (baryons and leptons), but not by bosons (photons and mesons).

peculiar velocity   The velocity of an object that is different from the average velocity of the objects in its surroundings.

period-luminosity relation   The observed relationship between the periods of pulsation and the absolute luminosities of Cepheid variable stars.

perturbation   A deviation from the normal, as in the orbital motion of an object or objects or in the distribution of material in a medium.

photo-ionization   The process involving the ionization of a gas by energetic photons.

photometry   The measurement of the brightness of an object.

photon   A quantum particle of electromagnetic radiation (light). The energy of a photon is proportional to the frequency of the corresponding light wave, and inversely proportional to the wavelength. The photon conveys the electromagnetic force.

photosphere   The layer of a star where the opacity is small enough that the light can freely escape. It is the surface we see.

planetary nebula   An ejected shell from a highly evolved star.

plasma   A gas consisting of ionized atoms and free electrons. The charges of these ions and electrons make the gas especially sensitive to magnetic fields and electromagnetic radiation (light).

plunging orbit   An elliptical orbit with a very large eccentricity, or an unbound hyperbolic orbit, that carries the object in an almost radial path to and from the center.

polycyclic aromatic hydrocarbons (PAHs)   Organic molecules that are based on the benzene ring molecule, C_6H_6, and are detected in circumstellar and interstellar environments.

Population I and II   The two main types of populations of stars in our Galaxy. Population I stars are young and Population II stars are old.

position angle   The orientation of a line in the plane of the sky measured counterclockwise with respect to north.

potential energy   The energy inherent in a configuration that has yet to be actualized.

proper motion   The apparent motion of a star in the sky with respect to background stars or galaxies, measured in arcseconds per year.

protogalaxies   Objects in the Universe that preceded the formation of galaxies.

proton   A positively charged subatomic particle that belongs to the hadron family of elementary particles and exists in atomic nuclei.

protostar   A star that has not yet reached its equilibrium state on the main sequence.

pulsar   A rotating magnetized neutron star that emits radio and optical light pulses in a manner similar to that of a lighthouse.

quantum   Pertaining to subatomic particles, where uncertainties in position, momentum, time, and energy are important, and where observed behavior is explained in terms of wave physics and probabilistic theories. Also pertaining to discrete allowed configurations of specified charge, mass, angular momentum, and energy—as in quanta of light (photons) and the quantized states of atoms.

quark   A subatomic particle of fractional charge whose various combinations make up the hadron family of particles, including protons, neutrons, and mesons. Quarks are found only in these combinations—never in isolation. They are especially subject to the strong force, but also respond to the weak, electromagnetic, and gravitational forces.

quasar   The brilliant, small source of light found in certain distant galaxies and resulting from the presence of a super-massive black hole in the galaxy's nucleus that is accreting its immediate surroundings.

radial velocity   The measured velocity along the line of sight.

radio galaxies Giant galaxies, usually involving active nuclei and/or tidal interactions, that emit strong radio-wavelength radiation.

reddening The selective absorption of shorter-wavelength light by the interstellar medium, resulting in the selective transmission of longer-wavelength (redder) light. The setting Sun is reddened by the Earth's atmosphere in a similar way.

red giant A star that has expanded and cooled following its main-sequence phase.

redshift The Doppler shifting of observed light to longer (redder) wavelengths, caused by recession of the source.

relativity The formulations of natural law that incorporate the constancy of the velocity of light in any reference frame (special relativity) and the relation between gravity and the curvature of space-time (general relativity).

relaxation The dynamical smoothing and rearranging of stars in a system caused by mutual gravitational interactions.

resolution The level of detail that can be discerned in an image, spectrum, or light curve.

rotation curve A plot of the velocity of rotation against the distance from the center, for instance, in a disk galaxy.

RR Lyrae variable star A short-period pulsating giant star, commonly found in old globular clusters, whose near-constant average brightness is used as a standard candle to gauge the distances of these clusters.

Schwarzschild radius The radius demarcating the event horizon of a spherical black hole, inside of which light or any other form of matter-energy is unable to escape.

self-gravitation The property of a body that is dominated by its own gravitational field, exclusive of outside influences.

Seyfert galaxies A class of galaxies with brilliant nuclei exhibiting strong and broad spectral-line emission.

shock ionization The process in which the atoms in a gas are collisionally ionized by the passage of a shock front that rapidly compresses and energizes the gas.

singularity The dimensionless point at the center of a black hole where all of the mass is located and the corresponding density is infinite.

Solar Neighborhood The immediate area around the Sun, including the nearest few thousand stars.

space-time An inclusive term that relates the three dimensions of space and one dimension of time in the presence of gravity and other energy fields. The general theory of relativity explicitly uses space-time in its formulations.

spectral energy distribution (SED) The relationship between the amount of energy emitted and the wavelength of the emitted light.

spectral line indices Measures of the strengths of spectral lines of certain elements in the spectrum of a star or galaxy.

spectral type A classification of the spectrum of a star according to which lines are prominent. A star's spectral type is related to the surface temperature and gravity of the star.

spectroscope An instrument for recording the spectra of objects.

spectrum The light of a source, spread out into its different wavelengths.

starburst A galaxy or portion thereof that is experiencing an energetic episode of enhanced star formation involving large numbers of hot massive stars and associated supernova activity.

stochastic Governed by randomly occurring processes.

Stromgren sphere An H II region of ionized gas around a hot star.

supercluster An unusually large or massive cluster of either stars or galaxies.

supergiant An evolved, massive star that is brighter than the more common giant stars.

supernova The explosive destruction of a massive star, following the implosion of its iron core to a neutron star or black hole. Alternatively, the explosive destruction of a white dwarf stellar remnant whose mass exceeds the Chandresekhar limit of 1.4 solar masses. A Supernova is observed as an extreme brightening of the original stellar source.

supernova remnant (SNR) The gaseous cloud of material ejected by the explosion of a supernova and/or of interstellar material that has been plowed up by the supernova's blast wave.

synchrotron radiation   Light emitted by charged particles moving at velocities close to that of light in the presence of a magnetic field. The radiation has a unique nonthermal spectrum and is highly polarized.

thermal   Pertaining to an ensemble of objects with a common temperature. Thermal radiation is that emitted by such an ensemble. Its spectrum is similar to that of an idealized black body.

tidal interactions   The effects of gravitational tides of one body on another where the tides involve differential gravitational forces across each of the interacting bodies.

vacuum   Empty space that is nonetheless imbued with latent energy. Quantum fluctuations of energy, time, and matter may characterize the vacuum.

velocity dispersion   The range in velocities seen, for example, in the stars occupying the center of a galaxy.

velocity of escape   The velocity that an object must have away from another body in order to become clear of the gravitational attraction of the other body.

virtual particle   An ephemeral particle that defies the dictum of matter-energy conservation, but whose appearance and disappearance occurs on timescales permitted by Heisenberg's uncertainty principle, thereby circumventing the violation. The higher the energy of the particle, the briefer the allowed timescale. Virtual particle–antiparticle pairs are thought to be continuously arising from the quantum vacuum.

weakly interacting massive particles (WIMPs)   Particles that may constitute a significant fraction of the dark matter in the Universe.

white dwarf star   The collapsed remnant core of an evolved star of intermediate mass that has exhausted its nuclear fuel. The core is made up of helium and carbon nuclei and is held up by the degenerate pressure exerted by charged electrons. The upper mass limit of a white dwarf is 1.4 solar masses, above which the core collapses to a neutron star or black hole, or explodes.

Wolf-Rayet star   An extremely hot, massive star that is ejecting gas shells at high velocity.

X-rays   High-energy electromagnetic radiation with wavelengths of about 10 Angstroms to 100 Angstroms—shorter than those of ultraviolet light but longer than those of gamma rays.

ILLUSTRATION

CREDITS

Abbreviations and Acronyms Used

AAO	Anglo-Australian Observatory
ACS	Advanced Camera for Surveys
APO	Apache Point Observatory
ARC	Astronomical Research Corporation
ASP	Astronomical Society of the Pacific
ATCA	Australia Telescope Compact Array
ATNF	Australia Telescope National Facility
AUI	Associated Universities Incorporated
AURA	Association of Universities for Research in Astronomy
BATSE	Burst and Transient Source Experiment
Caltech	California Institute of Technology
CfA	Harvard-Smithsonian Center for Astrophysics
CGRO	Compton Gamma-Ray Observatory
CITA	Canadian Institute for Theoretical Astrophysics
COBE	Cosmic Background Explorer
CTIO	Cerro-Tololo Inter-American Observatory
CXC	Chandra X-Ray Center
DIRBE	Diffuse Infrared Background Experiment
DMR	Differential Microwave Radiometer
ESA	European Space Agency
ESO	European Southern Observatory
FCRAO	Five College Radio Astronomy Observatory
FORS	FOcal Reducer Spectrograph
GSFC	Goddard Space Flight Center
HEASARC	High Energy Astrophysics Science Archive Research Center

HST	Hubble Space Telescope
IPAC	Infrared Processing and Analysis Center
IRAS	Infrared Astronomy Satellite
ISO	Infrared Space Observatory
JHU	The Johns Hopkins University
KPNO	Kitt Peak National Observatory
LO	Lick Observatory
MLO	Mount Laguna Observatory
MPIfR	Max Planck Institute for Radiophysics
NASA	National Aeronautics and Space Administration
NCSA	National Center for Supercomputing Applications
NOAO	National Optical Astronomy Observatories
NRAO	National Radio Astronomy Observatory
NRL	Naval Research Laboratory
NSF	National Science Foundation
SDSS	Sloan Digital Sky Survey
STScI	Space Telescope Science Institute
2MASS	Two Micron All Sky Survey
UCLA	University of California at Los Angeles
UCSB	University of California at Santa Barbara
UCSC	University of California at Santa Cruz
UCSD	University of California at San Diego
UIT	Ultraviolet Imaging Telescope
UIUC	University of Illinois at Urbana-Champaign
UMass	University of Massachusetts
VLA	Very Large Array
VLT	Very Large Telescope
WIYN	Wisconsin-Indiana-Yale-NOAO 3.5-meter Telescope

Figures

1.1 Radio: J. Condon et al., VLA, NRAO, AUI, NSF. Infrared: I. F. Mirabel, O. Laurent (Service d'Astrophysique, Commissariat a l'Énergie Atomique), et al., ISOCAM Team, ISO, ESA. Visible: VLT, ESO. X-ray: M. Karovska et al., Chandra X-Ray Observatory, CXC, CfA, NASA.

1.2 Top: Digitized Sky Survey, STScI, AURA, NASA, and SkyView Virtual Observatory, GSFC, NASA. Middle: J. Zabielski, A. Block, 0.4-meter telescope, KPNO, NOAO, AURA, NSF. Bottom: V. Andersen, 0.9-meter telescope, KPNO, NOAO, AURA, NSF.

1.3 Adapted from E. Hubble, *The Realm of the Nebulae* (New Haven: Yale University Press, 1936), p. 45.

1.4 Top: Digitized Sky Survey, STScI, AURA, NASA, and SkyView Virtual Observatory, GSFC, NASA. Middle and bottom: P. W. Hodge, *Galaxies* (Cambridge, Mass.: Harvard University Press, 1986), pp. 18, 19, 20, 24.

1.5 Adapted from G. de Vaucouleurs, in *Handbuch der Physik,* vol. 53, *Astrophysics IV: Stellar Systems,* ed. S. Flügge (Berlin: Springer-Verlag, 1959), p. 282.

1.6 Ibid., p. 288.

1.7   Images: Digitized Sky Survey, STScI, AURA, NASA, and SkyView Virtual Observatory, GSFC, NASA. Plot: Adapted from K. Freeman, in *Galaxies,* Saas-Fee Advanced Course 6, Swiss Society for Astrophysics and Astronomy (Sauverny, Switzerland: Geneva Observatory, 1976), p. 16

2.1   W. H. Waller.

2.2   W. H. Waller.

2.3   W. H. Waller.

3.1   Adapted from G. G. Fazio, in *Frontiers of Astrophysics,* ed. E. H. Avrett (Cambridge, Mass: Harvard University Press, 1976), p. 206; based on B. Neugebauer and E. E. Becklin, "The Brightest Infrared Sources," in *Scientific American,* 228 (1973), 28–40.

3.2   Adapted from L. S. Sparke and J. S. Gallagher, *Galaxies in the Universe* (Cambridge: Cambridge University Press, 2000), p. 5; courtesy of L. S. Sparke.

3.3   Top: From data in the HIPPARCOS catalogue as presented by R. F. Garrison in *Observers Handbook, 2001,* ed. R. Gupta, Royal Astronomical Society of Canada (Toronto: University of Toronto Press, 2001), p. 220. Bottom: From models in G. Bertelli et al., in the *Astronomy and Astrophysics Supplement,* 106 (1994), 275.

3.4   Adapted from H. Jahreiss and R. Wielen, in *ESA SP-402: Proceedings of the Hipparcos-Venice '97 Symposium,* ed. M. A. C. Perryman and P. L. Bernacca, 402 (1997), 675.

3.5   Adapted from L. S. Sparke and J. S. Gallagher, *Galaxies in the Universe* (Cambridge: Cambridge University Press, 2000), p. 205; courtesy of L. S. Sparke.

3.6   WIYN 3.5-meter telescope, copyright WIYN Consortium, Inc., all rights reserved, NOAO, AURA, NSF.

4.1   Reprinted by permission of the publishers from *Galaxies* by Paul W. Hodge, Cambridge, Mass.: Harvard University Press, p. 55; copyright © 1986 by the President and Fellows of Harvard College.

4.2   Ibid., p. 56.

4.3   Courtesy of Y. Sofue (Institute of Astronomy, University of Tokyo), with reference to Y. Sofue et al., in the *Astrophysical Journal,* 523 (1999), 136.

4.4   Courtesy of J. A. Tyson (Bell Laboratories, Lucent Technologies), with reference to J. A. Tyson, G. P. Kochanski, and I. P. Dell 'Antonio, in the *Astrophysical Journal Letters,* 498 (1998), 107.

4.5   CXC, Smithsonian Astrophysical Observatory, CfA, NASA.

5.1   Reprinted by permission of the publishers from *Galaxies* by Paul W. Hodge, Cambridge, Mass.: Harvard University Press, p. 30, 31; copyright © 1986 by the President and Fellows of Harvard College.

5.2   Ibid., pp. 33, 34.

6.1   Plaster cast of the cover of the sarcophagus of Nut, gift of Edward S. Harkness, 1914, accession number 14.7.1 b, all rights reserved, the Metropolitan Museum of Art, New York City, New York.

6.2   Top: J. Bieniasz, Griffith Observatory; courtesy of E. Krupp, Griffith Observatory. Middle: W. Herschel, *On the Construction of the Heavens,* in *The Scientific Papers of Sir William Herschel* (London: The Royal Society and the Royal Astronomical Society, 1912), p. 251. Bottom: Adapted from J. S. Plaskett, in *Journal of the Royal Astronomical Society of Canada,* 30 (1936), 153.

6.3   Top: Adapted from R. J. Maddalena, who used the SkyMap 3.0 desktop planetarium program. Bottom: Courtesy of R. J. Maddalena, 1.2-meter millimeter-wave telescope, Columbia University, with reference to R. J. Maddalena et al., in the *Astrophysical Journal,* 303 (1986), 375.

6.4 Digitized Sky Survey, STScI, AURA, NASA, and SkyView Virtual Observatory, GSFC, NASA.

6.5 Courtesy of H. Freudenreich (Raytheon, GSFC), DIRBE, COBE, NASA, with reference to H. Freudenreich in *American Scientist,* 87 (1999), 418.

6.6 B. P. Wakker (University of Wisconsin), I. Kallick (Possible Designs, Madison, Wisc.), STScI, AURA, NASA, with reference to H. Van Woerden et al., in *Nature,* 400 (1999), 138. Milky Way image: Lund Observatory, Sweden. High-velocity clouds: Dwingeloo Radio Astronomy Observatory, Netherlands.

6.7 Adapted from D. Clemens (Boston University), 14-meter millimeter-wave telescope, FCRAO, NSF, with reference to D. P. Clemens, D. B. Sanders, and N. Z. Scoville, in the *Astrophysical Journal,* 327 (1988), 139.

6.8 B. J. Bok and P. F. Bok, *The Milky Way,* 5th ed. (Cambridge, Mass.: Harvard University Press, 1981), p. 272; figure based on data from W. Becker and Th. Schmidt-Kaler.

6.9 Courtesy of H. Freudenreich (Raytheon, GSFC), DIRBE, COBE, NASA, with reference to H. Freudenreich in *American Scientist,* 87 (1999), 418.

7.1 R. Smith, CTIO, NOAO, AURA, NSF.

7.2 Left: CTIO, NOAO, AURA, NSF. Top right: 4-meter Blanco telescope, CTIO, NOAO, AURA, NSF. Middle right: M. Romaniello (ESO), HST, STScI, NASA, ESA. Bottom right: HST, STScI, AURA, NASA.

7.3 Adapted from P. W. Hodge, in *Annual Reviews of Astronomy and Astrophysics,* 27 (1989), 139.

7.4 W. H. Waller and J. D. Offenberg, *Beyond the Blue: Greatest Hits of the Ultraviolet Imaging Telescope,* slide set copyright 1994 ASP; based on data from UIT, HST, NASA.

7.5 Reprinted by permission of the publishers from *Galaxies* by Harlow Shapley, Cambridge, Mass.: Harvard University Press; copyright © 1943, 1961, 1972 by the President and Fellows of harvard College. Top: Adapted from p. 59. Bottom: Adapted from p. 61; from H. S. Leavitt, *Harvard College Observatory Circular* no. 173 (1912), p. 1.

7.6 Top: Courtesy of C. Bruens (University of Bonn), 64-meter Parkes radio telescope, ATNF. Bottom: Courtesy of S. Kim, L. Staveley-Smith, ATCA, ATNF, with reference to S. Kim, L. Staveley-Smith, et al., in the *Astrophysical Journal,* 503 (1998), 674.

8.1 Courtesy of E. K. Grebel, Max Planck Institute for Astronomy, with reference to E. K. Grebel, in *The Stellar Content of the Local Group: IAU Symposium 192,* ed. P. Whitelock and R. Cannon (San Francisco: ASP, 1999), p. 17.

8.2 Adapted from P. W. Hodge, in *Annual Reviews of Astronomy and Astrophysics,* 27 (1989), 139.

8.3 Top: P. W. Hodge, 1.2-meter Schmidt telescope, Palomar Observatory. Bottom: Digitized Sky Survey, STScI, AURA, NASA, and SkyView Virtual Observatory, GSFC, NASA.

8.4 Digitized Sky Survey, STScI, AURA, NASA, and SkyView Virtual Observatory, GSFC, NASA.

8.5 United Kingdom Schmidt Telescope (UKST), Siding Spring Observatory, AAO, New South Wales, Australia, and the Royal Observatory, Edinburgh. The UK Schmidt

Telescope was operated by the Royal Observatory Edinburgh, with funding from the UK Science and Engineering Research Council, until late June 1988 and thereafter by the AAO. Original plate material is copyright the Royal Observatory Edinburgh and the AAO.

9.1 Top: Adapted from P. W. Hodge, *Galaxies* (Cambridge, Mass.: Harvard University Press, 1986), p. 115. Bottom: P. W. Hodge, 4-meter telescope, KPNO, NOAO, AURA, NSF.

9.2 M. Haas, D. Lemke, M. Stickel, H. Hippelein, et al., ISOPHOT Camera Team, ISO, ESA.

9.3 Left: 1.2 meter Schmidt telescope, Palomar Observatory. Right: H. Yang (University of Illinois), HST, STScI, AURA, NASA.

9.4 Courtesy of D. S. Adler (STScI), VLA, NRAO, AUI, NSF, with reference to D. S. Adler and D. J. Westpfahl, in the *Astrophysical Journal*, 111 (1996), 735.

10.1 Top: F. Zwicky, in *Physics Today*, 6 (1953), 7; reprinted by F. Schweizer in *Galaxies: Interactions and Induced Star Formation* (Berlin: Springer-Verlag, 1998). Bottom: Adapted from J. E. Hibbard (NRAO), KPNO and CTIO telescopes, NOAO, AURA, NSF, with reference to J. E. Hibbard and H. H. van Gorkom, in the *Astronomical Journal*, 111 (1996), 655.

10.2 Top: Digitized Sky Survey, STScI, AURA, NASA, and SkyView Virtual Observatory, GSFC, NASA. Bottom: W. H. Waller, UIT, NASA, with reference to W. H. Waller et al., in the *Astrophysical Journal*, 481 (1997), 169.

10.3 Left: Digitized Sky Survey, STScI, AURA, NASA. Right: VLA, NRAO, AUI, NSF, with reference to M. S. Yun, P. T. P. Ho, and K. Y. Lo, in *Nature*, 372 (1994), 530.

11.1 K. D. Borne (NASA's GSFC), HST, STScI, AURA, NASA, with reference to K. D. Borne et al., in the *Astrophysical Journal Letters*, 529 (1999), L77.

11.2 Courtesy of C. Carilli and G. Taylor (NRAO), VLA, NRAO, AUI, NSF.

11.3 Reprinted by permission of the publishers from *Galaxies* by Paul W. Hodge, Cambridge, Mass.: Harvard University Press, p. 157; copyright © 1986 by the President and Fellows of Harvard College.

11.4 J. N. Bahcall (Institute for Advanced Studies, Princeton), Mike Disney (University of Wales), HST, STScI, AURA, NASA.

11.5 Adapted from C. M. Urry (Yale University) and P. Padovani (STScI), HEASARC, GSFC, NASA.

12.1 Reprinted by permission of the publishers from *Galaxies* by Paul W. Hodge, Cambridge, Mass.: Harvard University Press, p. 132; copyright © 1986 by the President and Fellows of Harvard College.

12.2 Ibid., p. 137.

12.3 Adapted from W. L. Freedman et al., in the *Astrophysical Journal*, 553 (2001), 47; courtesy of W. L. Freedman.

13.1 KPNO 4-meter Mayall telescope, NOAO, AURA, NSF.

13.2 Courtesy of M. Hudson (University of Waterloo), with reference to M. Hudson, in *Monthly Notices of the Royal Astronomical Society*, 265 (1993), 42.

13.3 Top: Courtesy of J. Huchra, CfA2 Redshift Survey, Harvard-Smithsonian Center for Astrophysics, with reference to M. Geller and J. Huchra, in *Science*, 246 (1989), 897. Bottom: Las Campanas Redshift Survey, Carnegie Institution's Las Campanas Observatory, with reference to S. Shectman et al., in *Astrophysical Journal*, 470 (1996), 172.

14.1 Adapted from J. Bennett, M. Donahue, N. Schneider, and M. Voit, *Cosmic Perspectives,* 2nd ed. (Reading, Mass.: Addison-Wesley, 2002), p. 629.

14.2 Adapted from N. F. Comins and W. J. Kaufman III, *Discovering the Universe,* 4th ed. (New York: W. H. Freeman, 1997), p. 377.

14.3 Adapted from C. J. Hogan, R. P. Kirshner, and N. B. Suntzeff, in *Scientific American,* January 1999, p. 93; courtesy of the High-Z Supernova Search Team.

15.1 Courtesy of C. Bennett, DMR, COBE, GSFC, NASA.

15.2 Adapted from R. V. Wagoner, in the *Astrophysical Journal,* 179 (1973), 343.

16.1 M. Hauser (STScI), DIRBE, COBE, GSFC, NASA.

16.2 BATSE, CGRO, HEASARC, GSFC, NASA.

16.3 R. Griffiths (The Johns Hopkins Institute), Medium Deep Survey, HST, STScI, AURA, NASA.

16.4 HST, STScI, AURA, NASA.

16.5 Courtesy of W. C. Keel. Top: HST, STScI, AURA, NASA. Bottom: Keck I 10-meter telescope, William Keck Observatory, Hawaii.

Plates

1 Top left: KPNO, NOAO, AURA, NSF. Top right: KPNO, NOAO, AURA, NSF. Middle left: KPNO, NOAO, AURA, NSF. Middle right: SDSS, APO, ARC. Bottom left: H. Mathis, N. A. Sharp, KPNO, NOAO, AURA, NSF. Bottom right: NOAO, AURA, NSF.

2 Top left: M. Chase, A. Block, KPNO, NOAO, AURA, NSF. Top right: KPNO, NOAO, AURA, NSF. Middle left: KPNO, NOAO, AURA, NSF. Middle right: A. Block, KPNO, NOAO, AURA, NSF. Bottom left: D. Seibel, A. Block, KPNO, NOAO, AURA, NSF. Bottom right: Copyright 1992–2002 D. Malin, AAO, all rights reserved.

3 Top left: A. Block, KPNO, NOAO, AURA, NSF. Top right: CTIO, NOAO, AURA, NSF. Middle left: CTIO, NOAO, AURA, NSF. Middle right: A. Block, KPNO, NOAO, AURA, NSF. Bottom left: H. Ford (JHU), G. Illingworth (UCSC, LO), M. Clampin (STScI), G. Hartig (STScI), the ACS Science Team, HST, NASA, ESA. Bottom right: Hubble Heritage Team, HST, STScI, AURA, NASA.

4 W. H. Waller and J. D. Offenberg, *Beyond the Blue: Greatest Hits of the Ultraviolet Imaging Telescope,* slide set copyright 1994 ASP. Top: UIT, NASA. Bottom: MLO and KPNO, NOAO, AURA, NSF.

5 Top: UIT, NASA. Bottom: 0.9-meter telescope, KPNO, NOAO, AURA, NSF, with reference to W. H. Waller et al., in the *Astrophysical Journal,* 121 (2001), 1395.

6 Top: Copyright 2000 A. Mellinger, University of Potsdam, all rights reserved. Bottom: J. Carpenter, M. Skrutskie, R. Hurt, 2MASS, UMass, NSF, IPAC, Caltech, NASA.

7 Top: E. L. Wright (UCLA), DIRBE, COBE, GSFC, NASA. Bottom: C. Haslam et al., Jodrell Bank, MPIfR, Parkes Observatory, SkyView Virtual Observatory, GSFC, NASA.

8 IRAS, IPAC, NASA, ESA.

9 Top: ISO, ESA. Bottom: VLT, ESO.

10 Top: C. R. O'Dell and S. K. Wong (Rice University), HST, STScI, AURA, NASA. Bottom: J. Bally (University of Colorado), D. Devine, and R. Southerland (CITA), HST, STScI, AURA, NASA.

11 Top left: Copyright 1997 P. Berlind and P. Challis, 1.2-meter telescope, Whipple Observatory, CfA. Top right: J. Morse (University of Colorado), K. Davidson (University of Minnesota), HST, STScI, AURA, NASA. Bottom: WIYN 3.5-meter telescope, copyright 2000 WIYN Consortium, Inc., all rights reserved, NOAO, AURA, NSF.

12 W. H. Waller (Tufts University), F. Boulanger (Institut d'Astrophysics), F. Varosi (University of Florida), IRAS, IPAC, NASA, ESA, and SkyView Virtual Observatory, GSFC, NASA.

13 Top: N. E. Kassim, D. S. Briggs, T. J. Lazio, T. N. LaRosa, J. Imamura, and S. D. Hyman (NRL), A. Pedlar, K. Anantharamiah, M. Goss, and R. Ekers (VLA, NRAO, AUI, NSF). Bottom left: F. Yusef-Zadeh, M. R. Morris, D. R. Chance, VLA, NRAO, AUI, NSF. Bottom right: W. M. Goss, R. D. Ekers, J. H. van Gorkom, U. J. Schwarz, VLA, NRAO, AUI, NSF.

14 The 4-meter Blanco telescope, CTIO, NOAO, AURA, NSF.

15 The 4-meter Blanco telescope, CTIO, NOAO, AURA, NSF.

16 T. A. Rector and B. A. Wolpa, 0.9-meter telescope, KPNO, NOAO, AURA, NSF.

17 T. A. Rector and M. Hanna, 0.9-meter telescope, KPNO, NOAO, AURA, NSF.

18 W. H. Waller and J. D. Offenberg, *Beyond the Blue: Greatest Hits from the Ultraviolet Imaging Telescope,* slide set copyright 1994 ASP; based on data from UIT, NASA, and KPNO, NOAO, AURA, NSF.

19 FORS team, 8.2-meter ANTU telescope, VLT, ESO.

20 Top: G. F. Benedict, McDonald Observatory, University of Texas, HST, STScI, AURA, NASA. Bottom: Hubble Heritage Team, HST, STScI, AURA, NASA.

21 Top: Composite of images at the bottom, CXC, CfA, NASA. Bottom: X-ray: M. Karovska et al., Chandra X-Ray Observatory, CXC, CfA, NASA. Optical: United Kingdom Schmidt telescope, Digitized Sky Survey, STScI. Radio continuum: J. Condon et al., VLA, NRAO, AUI, NSF. Radio 21-centimeter: L. Schiminovich et al., VLA, NRAO, AUI, NSF.

22 T. A. Rector and Monica Ramirez, 0.9-meter telescope, KPNO, NOAO, AURA, NSF.

23 B. Whitmore, HST, STScI, AURA, NASA.

24 K. D. Borne, HST, STScI, AURA, NASA.

25 Top: C. Martin (UCSB), 2.1-meter telescope, KPNO, NOAO, AURA, NSF. Bottom: C. Martin et al., Chandra X-ray Observatory, CXC, CfA, NASA, and KPNO, NOAO, AURA, NSF.

26 Subaru Telescope, copyright 2000 the National Astronomical Observatory of Japan, all rights reserved.

27 Top: Hubble Heritage Team, HST, STScI, AURA, NASA. Bottom: Faint Object Spectrograph, HST, STScI, AURA, NASA.

28 Top: O. Lopez-Cruz and I. Shelton, 0.9-meter telescope, KPNO, NOAO, AURA, NSF. Bottom: W. Baum (University of Washington), HST, STScI, AURA, NASA.

29 Courtesy of J. A. Tyson (Bell Laboratories, Lucent Technologies), HST, STScI,

AURA, NASA, with reference to J. A. Tyson, G. P. Kochanski, and I. P. Dell 'Antonio, in the *Astrophysical Journal Letters,* 498 (1998), 107.

30   R. Williams and the Hubble Deep Field Team, HST, STScI, AURA, NASA.

31   Courtesy of G. L. Bryan and M. L. Norman (UCSD), NCSA, UIUC, NSF, with reference to G. L. Bryan and M. L. Norman in the *Astrophysical Journal,* 495 (1998), 80.

32   DMR, COBE, GSFC, NASA.

INDEX